The Canadian Institute of
Resources Law

AF712712

THE INTERNATIONAL LEGAL CONTEXT OF PETROLEUM OPERATIONS IN CANADIAN ARCTIC WATERS

Property of
I. N. McKINNON MEMORIAL LIBRARY

Ian Townsend Gault

Working Paper 4
Canadian Continental Shelf Law 2

July 1983

Faculty of Law. The University of Calgary, 2500 University Drive N.W., Calgary, Alberta. T2N 1N4 Tel. (403) 282-9197

Canadian Cataloguing in Publication Data

Townsend Gault, Ian, 1952-
The International Legal Context of Petroleum Operations in Canadian Arctic Waters

(Canadian continental shelf law ; 2) (Working paper ; 4)
Includes bibliographical references.
ISBN 0-919269-10-9

1. Offshore oil industry - Law and legislation.
2. Offshore gas industry - Law and legislation.
3. Continental shelf - Arctic regions. 4. Arctic regions - International status. I. Canadian Institute of Resources Law. II. Title. III. Series. IV. Series: Working paper (Canadian Institute of Resources Law) ; 4.

K3918.T68 341.7'62 C83-091288-6

Copyright 1982 by Canadian Institute of Resources Law
The University of Calgary
All Rights Reserved
Printed in Canada

THE CANADIAN INSTITUTE OF RESOURCES LAW

The Canadian Institute of Resources Law was incorporated in 1979 with a mandate to examine the legal aspects of both renewable and non-renewable resources. Its work falls into three inter-related areas: research, education, and publication.

The Institute has engaged in a wide variety of research projects, including studies on environmental regulation of oil and gas projects; compensation schemes for damages arising from offshore petroleum licensing systems; Canadian continental shelf law; uncertainty in the regulation of mining projects; and forestry and water law.

The Institute will also consider participation in multidisciplinary research and in joint projects with other institutions. The Institute does not undertake research for individual clients unless, in the opinion of the Board of Directors, those problems raise broader issues related to the Institute's objectives and aims. The Institute's integrity is secured by its retention in all cases of the right to publish the results of its research.

The education function of the Institute is pursued by sponsoring conferences and short courses on particular topical aspects of Resources Law, and through teaching in the Faculty of Law at The University of Calgary.

The major publication of the Institute is its ongoing looseleaf service, the Canada Energy Law Service, published in association with Richard de Boo Ltd. The results of other Institute research are published as monographs and working papers. Manuscripts submitted by outside authors will also be considered. The Institute publishes a newsletter, Resources, on an occasional basis.

The Institute is supported by the Alberta Law Foundation, the Government of Alberta, the Government of Canada, the Foundation for Legal Research, the Committee on Energy and Natural Resources of the International Bar Association, and the private sector. The members of the Board of Directors are appointed by the Faculty of Law at The University of Calgary, the President of The University of Calgary, the President of the Law Society of Alberta, the President of the Canadian Petroleum Law Foundation, and the Dean of Law at The University of Alberta. Additional members of the Board are elected by the appointed Directors.

All enquiries should be addressed to:

The Executive Director
Canadian Institute of Resources Law
Rm 430, BioSciences Bldg.
Faculty of Law
The University of Calgary
Calgary, Alberta, Canada T2N 1N4

Telephone: (403)282-9197
Telex: 03-821545

L'INSTITUT CANADIEN DU DROIT DES RESSOURCES

L'Institut Canadien du Droit des Ressources fut incorporé en 1979 dans le but d'étudier les aspects légaux touchant les ressources renouvelables et rien-renouvelables. Son travail s'étend sur trois champs d'action inter -dépendants qui sont: la recherche, l'éducation et la publication.

L'Institut s'est impliqué dans une grande variété de projets de recherche dant des etudes sur la reglementation vis-à-vis de l'environnement des projets de gaz naturel et de pétrole; des systèmes de compensation pour les dammages caussés par les opérations pétrolières au large des côtes maritimes; des systèmes de permis pétroliers; le droit Canadien en ce qui affectes pe plateau continental; les doutes dans la reglementation des projets miniers; le droit pour ce qui est des forêts et des eaux.

L'Institut prendra en considération les demandes de participation dans des projet de recherche inter-disciplinaires et en collaboration avec d'autres institutions. L'Institut n'entreprend pas de recherches pour des clients particuliers à moins que, selan l'opinion du comité de direction, leurs problèmes ne soulèvent des questions de plus ample portée qui sont reliées aux buts et aux objectifs de l'Institut. L'intégrité de celui-ci est assuré par le droit de publication des résultats de ses recherches qu'il se réserve dans tous les cas.

La fonction éducative de l'Institut se poursuit tant par les conférences qu'il parraine que par les minicours sur des aspects particuliers du droit des ressources de même que par l'enseignement à la faculté de Droit de l'Université de Calgary.

La publication principale de l'Institut est son service continu de feuilles volantes, "Canada Energy Law Service", publié en collaboration avec Richard De Boo Ltée. Le résultats d'autres recherches de l'Institut sont publié en tant que monographes et travaux d'utilisation courante. Les manuscrits soumis par des auteurs non-associés a l'Institut seront aussi pris en considération. L'Institut publie, a l'occasion, une lettre circulaire: "Resources".

L'Institut es appuyé par le gouvernement de l'Alberta, le gouvernement Canadien, la Fondation pour la Recherche Legale, le Comité sur l'Energie et les Ressources Naturelles de l'Association Internationnale du Barreau de même que par le secteur privé. Les directeurs sont nommés par la faculté de Droit de l'Université de Calgary, le président de la Société de Droit de l'Alberta, le président de la Fondation Canadienne du Droit de l'Alberta, le président de la Fondation Canadienne du Droit Pétrolier et le doyen de la faculté de Droit de l'Université de l'Alberta. Des membres supplementaires du comité sont élus par les directeurs désignés.

Toute demande doit s'addresser à:

Le directeur exécutif
Institut Canadien du Droit des Ressources
Faculté de Droit
Université de Calgary
Calgary, Alberta, Canada
T2N 1N4

THE CANADIAN CONTINENTAL SHELF LAW PROJECT

This is the second in the Canadian Continental Shelf Law Project series of Working Papers to be published by the Canadian Institute of Resources Law. The Project, inaugurated in 1981, comprises an in-depth examination and analysis of Canadian law governing the conduct of offshore exploration, development, and transportation of hydrocarbons from the continental shelf. See "Publications" for information on Working Paper 1, Petroleum Operations on the Canadian Continental Margin - The Legal Issues in a Modern Perspective.

The areas for examination include the applicable systems for regulating oil and gas exploration and production, the international legal background, and the domestic Canadian jurisdictional, regulatory, and policy framework within which the legislation has been drafted and is being administered. Other legal issues arising in this context include the law applicable to the various offshore facilities used by the industry in the course of its operations, environmental law, admiralty law, and onshore planning law.

The Project is under the direction of Ian Townsend Gault, Research Associate at the Institute. The series of Working Papers will be published as work proceeds on a comprehensive treatise. Further details of the Project are available from the Project Director, who would be pleased to receive comments on this paper and on the research programme itself.

PROJET DE DROIT SUR LE PLATEAU CONTINENTAL CANADIEN

Ceci est le second d'une série de travaux de recherche concernant le Projet de loi sur le plateau continental canadien. Ce Projet, inauguré en 1981, comprend un examen detaillé de même qu'une analyse du droit canadien regissant l'exploration outre-cote, le développement et le transport des hydrocarbures en provenance du plateau continental. Voir "Publications" pour de plus amples information, sur le travail de recherche 1, Petroleum Operations on the Canadian Continental Margin - The Legal Issues in a Modern Perspective.

On examinera les moyens applicables pour contrôller l'exploration et la production des hydrocarbures. On se penchera aussi sur le contexte légal international, sur le cadre canadien en matière de juridiction, de contrôle et des lignes de conduite qui sont à la base des lois canadiennes. Ce champ d'ètude embrassera les autres problèmes légaux propres a ce secteur d'activité: les problemes que posent les installations qu'utilise l'industrie pétrolière, le droit de l'environnement, le droit d'amirauté de même que les lois régissant la planification générale.

Ian Townsend Gault, associé en recherche à l'Institut dirige le Projet. Tout au long du Projet de recherche qui se terminera par la publication d'un ouvrage complet on fera paraitre des documents de travail. Le directeur du Projet se met à la disposition de quicontre désire des reseignements ou voudrait lui adresser des commentaires ou sur le document de travail ou sur le Projet lui meme.

PUBLICATIONS

Following is a complete list of CIRL publications currently available.

Resources Law Bibliography. 1980. ISBN 0-919269-01-X 537 p.	$19.95
Environmental Regulation - Its Impact on Major Oil and Gas Projects: Oil Sands and Arctic, by C.D. Hunt and A.R. Lucas, 1980. ISBN 0-919269-00-1. 168 p.	$10.95
A Guide to Appearing before the Surface Rights Board of Alberta, by L.D. Ridsdel and R.J. Bennett. Working Paper 1, 1982. ISBN 0-919269-04-4. 70 p.	$5.00
Environmental Law in the 1980s: A New Beginning, Proceedings of a Colloquium, The Banff Centre, November 27-29, 1981, Peter Z.R. Finkle and Alastair R. Lucas, eds. Proceedings 1. ISBN 0-919269-05-2. 233 p.	$13.50
Petroleum Operations on the Canadian Continental Margin, by Ian Townsend Gault. Canadian Continental Shelf Law 1, Working Paper 2, 1983. ISBN 0-919269-05-2. 120 p.	$8.00
Acid Precipitation in North America: The Case For Transboundary Cooperation, by Douglas M. Johnston and Peter Finkle. 1983. ISNB 0-919269-02-8. 70 p.	$8.00

Please add: 10% for postage and handling (50% outside North America).

OTHER PUBLICATIONS

Canada Energy Law Service, ISBN 0-88820-108-7, available from

Richard de Boo Ltd.
81 Curlew Drive
Don Mills, Ontario, Canada
M3A 3P7

Resources - The Newsletter of the Canadian Institute of Resources Law, ISSN 0714-5918 — free

ACKNOWLEDGEMENTS

This is an expanded version of the report, "Petroleum Operations in Arctic Waters - Canada's Claims in an International Context", prepared by Mr. Rowland J. Harrison, formerly Executive Director of the Canadian Institute of Resources Law, and myself, commissioned by Dome Petroleum Limited, Esso Resources Canada Limited, and Gulf Canada Resources Inc., in connection with the preparation of an Environmental Impact Statement for hydrocarbon production in the Beaufort Sea - Mackenzie Delta area. The present version of the paper addresses issues not relevant in the context of the original report.

I would like once again to thank those who assisted in the preparation of the original report by sharing material or making time for interviews: Professor Donat Pharand, Faculty of Law, University of Ottawa; Dr. Harriet Critchley, Program Director, Northern Political Studies Program, Arctic Institute of North America, The University of Calgary; Mr. Allen Milne, formerly Manager, Environmental Impact Assessment and Mr. Murray Smith, Counsel, Dome Petroleum Limited; Professor Alastair Lucas, then Associate Executive Director, and Mr. Nigel Bankes, Research Associate, Canadian Institute of Resources Law.

Mr. Bill Matheson of the Department of Communications Media, The University of Calgary, executed the graphics. My thanks also to Mr. Don Sanders of The University of Calgary Faculty of Law Library for bibliographical services in connection with the original work, Mr. Murray Samuel and Mr. Christian Yoder, Research Assistants, and to Ms. Shirley Babcock and Ms. Evangeline Case of the Institute, for respectively typing and proof-reading the manuscript.

Mr. Harrison's substantial contribution to the present work was by way of his labours on the original report; responsibility for any errors or omissions herein are, of course, mine alone. I have attempted to state the law as of March 31, 1983.

Ian Townsend Gault
March 31, 1983
Calgary

TABLE OF CONTENTS

ABSTRACT

This paper identifies Canada's claims to jurisdiction in Arctic waters for the purpose of the exploration, production, and transportation of hydrocarbons from the continental shelf. Those claims are then placed in the context of the rights enjoyed by Canada at international law.

In Part I the development of the law of the sea and the doctrine of the continental shelf is traced. Petroleum operations in Canadian Arctic waters, and the domestic legal regime applicable thereto, are also outlined. In Part II a number of aspects of the domestic legal regime are examined - territorial sea and fishing zones, hydrocarbon resource exploration and production, environmental protection, transportation of oil and gas, and the construction of offshore facilities.

Part III analyzes contemporary international legal norms with respect to offshore jurisdictional zones, including the doctrine of the continental shelf, environmental protection, transportation of oil and gas, the construction of offshore facilities, and the legal status of Arctic waters, especially the Northwest Passage and the waters of the Canadian Arctic archipelago. The status of the sector theory and offshore boundary delimitation issues in the Canadian Arctic are also examined. Some conclusions and comments on the state of the legal regime, international and domestic, are offered in Part IV.

RESUME

Ce document identifie les réclamations du Canada à la juridiction des eaux de l'Arctique dans le but d'explorer, de produire et de transporter des hydrocarbures hors du plateau continental. Ces réclamations sont dès lors placées dans le contexte des droits possédés par le Canada pour la loi internationale.

Dans la partie I, le développement de la loi maritime et la doctrine du plateau continental sont tracés. Les opérations pétrolières des eaux Canadiennes de l'Arctique et le régime national légal applicable à cela sont aussi soulignés. Dans la partie II, un nombre d'aspects concernant le régime national légal sont examinés - le territoire maritime et les zones de pêche, l'exploration et la production des ressources hydrocarbures, la protection de l'environnement, le transport de l'huile et du gaz, et la construction d'installations en mer.

La partie III analyse les normes contemporaines internationales légales en respect aux zones juridictionnelles en mer, incluant la doctrine du plateau continental, la protection de l'environnement, le transport de l'huile et du gaz, la construction d'installations en mer, le statut légal des eaux de l'Arctique, spécialement le passage du Nord-Ouest et les eaux de l'archipel Canadien de l'Arctique. Le statut de la théorie du secteur et les questions de la délimitation de la frontière en mer sont aussi examinés. Quelques conclusions et commentaires sur l'état du régime légal, international et national sont offerts à la partie IV.

TABLE OF FIGURES

ACRONYMS AND ABBREVIATIONS

COGLA	Canada Oil and Gas Lands Administration
CRISTAL	Contract Regarding an Interim Supplement to Tanker Liability for Oil Pollution
DIAND	Department of Indian Affairs and Northern Development
EIS	Environmental Impact Statement
EM & R	Department of Energy, Mines and Resources
EARP	Environmental Assessment Review Process
FEARO	Federal Environmental Assessment Review Office
TOVALOP	Tanker Owners Voluntary Agreement Concerning Liability for Oil Pollution
UKOOA	United Kingdom Offshore Operators Association
UNCLOS III	Third United Nations Conference on the Law of the Sea

INTRODUCTION

This paper examines Canada's claims to jurisdiction for the purpose of regulating oil and gas activities in Arctic waters, and sets those claims in the context of the developing international legal regime pertaining to offshore petroleum operations. The focus is primarily legal, but the paper has been written so as to render it accessible to lawyers and non-lawyers alike. A glossary of legal terms and phrases is to be found in the Appendix.

INTERNATIONAL BACKGROUND

After almost a century of offshore oil and gas activities, it remains true that on the continental shelf the petroleum industry operates at the boundaries not only of technology, but also of law. It is therefore not altogether surprising that the development of domestic law which governs the conduct of offshore operations has been somewhat unsystematic. The nature of the demands which were to be made on the legal system were not always foreseen, and was not always foreseeable. It will, however, be argued in the course of this paper that remedial measures have not been implemented in Canadian law even when the need has been amply demonstrated in either the Canadian or foreign context.

Petroleum operations in the Canadian offshore are continuing to expand in response to government policy and incentives programs.[1] Much can be learned from the experiences of the United States, Australia, the United Kingdom, Norway, Nigeria, and other countries where offshore activities are well established.[2] Canada is almost alone in promoting resource exploration and development in the Arctic, where a new series of problems (jurisdictional, environmental, and technical) must be solved.

1. Government policy with respect to "frontier" development was set in: Canada. Dept. of Energy, Mines and Resources. Energy Policy Sector. The National Energy Program 1980, Report EP 80-4E [hereinafter NEP]. The latest federal incentives scheme - the Petroleum Incentives Program (PIP) - was implemented by the Petroleum Incentives Program Act (S.C. 1980-81-82-83, c. 107) and regulations made thereunder (the Petroleum Incentives Program Regulations, SOR/82-666).
2. Some 80 countries are currently engaged in continental shelf exploration; less than 30 are producing petroleum from the offshore.

Petroleum has been in production from the offshore since 1896, when wells were drilled from wooden platforms built from the shore near Santa Barbara, California. This initial experience enabled the petroleum industry to further develop its engineering techniques in the succeeding decades, most noticeably in Lake Maracaibo, Venezuela, in the late 1920s, and in the Caspian Sea. The most intense offshore petroleum producing effort anywhere in the world today is probably taking place in the North Sea area. Offshore development here, as elsewhere, has been prompted by an increased desire to secure domestic supplies of oil from all available sources. The legal and engineering background to offshore activity can be traced to the oil industry's experiences in the Gulf of Mexico following the second world war, experiences which built on the advances made between the wars. Without that experience, development of the North Sea, for example, would have been very much slower. Much of the North Sea experience is clearly applicable to other offshore areas, including the Canadian continental margin south of the 60th parallel of latitude. In the Arctic, however, industry and government are confronted with a new and different set of circumstances: technical, legal, and political.

The intensive exploration and production of the mineral resources of the continental shelf is, therefore, quite a recent development,[3] but one which has imposed and continues to impose unprecedented demands on the world's legal systems, domestic and international.[4]

---

3. The first unambiguous claim to exercise exclusive jurisdiction over the natural resources of the continental shelf was made by the U.S. in the so-called Truman Proclamation: Proclamation by the President of the United States, Department of State Bulletin #327, September 30, 1945.
4. Some of the demands which have already been made, and others which will be made, on the Canadian legal system are examined in the course of this paper. The North Sea states have already had to meet a number of such problems, e.g., the legal regime developed for joint production of oil and gas fields lying across the U.K./Norway continental shelf boundary. See Ian Gault, "The Frigg Gas Field", [1979] Marine Policy 302. The legal regime governing the exploitation of these fields is still developing. There have also been problems with court jurisdiction over offshore activities, and with certain aspects of the application of civil law; see, for example the recent English case of Clark (Inspector of Taxes) v. Oceanic Contractors Ltd., [1982] S.T.C. 66 (C.A.), reversed on appeal by the House of Lords ([1983] 2 W.L.R. 94). The point is that domestic law in the North Sea countries remains in a state of flux almost two decades after exploration commenced in that area.

The major uses of the world's oceans were, and continue to be, trade and fishing. The principles of the law of the sea were developed three centuries ago by European coastal states to ensure equal access to the seas for these purposes. The debate between those jurists who espoused the doctrine of open seas (and whose writings reflected the view of their governments), and their opponents, who favoured the alternative of closed seas for purposes such as defence and protectionism, was won by the former. Unimpeded access to all of the world's oceans was adjudged to be in the best interests of the users of the seas. In time, the suggestion that the high seas should be closed to either shipping or fishing was anathema, and the practices of the traditional sea users - distant water fishing fleets, for example - developed accordingly.[5]

The international legal principle that emerged in these circumstances was embodied in the principle of the "freedom of the seas". In international law, the high seas were not subject to the control or jurisdiction of any state.[6] National jurisdiction was restricted to a narrow band immediately adjacent to the coastline, known as the territorial sea, and historically accepted as having a width of one marine league or three nautical miles.[7]

In recent decades, however, the world's coastal nations have demanded more extensive jurisdiction over adjacent areas, for a number of reasons. Many states have, for example, denied foreign fishing fleets access to coastal fisheries in an

---

5. For a comprehensive treatment of the history of this area of the law see R.P. Anand, Origin and Development of the Law of the Sea, The Hague, Martinus Nijhoff Publishers, 1983. Professor Kunz, as late as 1956, saw the development of oil and gas activities on the continental shelf as a threat to the freedom of the seas. See Josef L. Kunz, "Continental Shelf and International Law: Confusion and Abuse", 50 A.J.I.L. 828 (1956).
6. Convention on the High Seas, April 29, 1958, in U.N. Doc. A/CONF 13/L.53, Art. 2.
7. Convention on the Territorial Sea and Contiguous Zone, April 29, 1958, in U.N. DOC. A/CONF 13/L.52 [hereinafter Territorial Sea Convention], Art. 1. Although this Convention did not set a limit to the territorial sea, the then accepted limit was three nautical miles. See Bernard G. Heinzen, "The Three Mile Limit: Preserving the Freedom of the Seas", 11 Stan. L. Rev. 597 (1959). Claims of up to 12 nautical miles are now generally accepted.

attempt to conserve stocks and protect local interests.[8] But the most important of these reasons in the present context are: first, controlling exploration for and exploitation of the mineral resources of the continental shelf, that area being regarded as the natural prolongation of the landmass appertaining to the coastal state;[9] and second, claims to increased jurisdiction in the offshore to mitigate or deal with pollution disasters. This claim has been justified by states on the grounds that the consequences of such pollution are frequently disasterous for the adjacent coastal nation which will often be unconnected to either ship or cargo. These claims have a special significance in the Arctic, which combines a vast potential for oil and gas production with a sensitive and hostile natural environment. International law has attempted to meet these needs by first developing and then refining a regime for oil and gas operations on the continental shelf.[10]

Exploration, production, and transportation of offshore resources continues to place additional strains on the existing legal regime, international and domestic. One source of these strains and stresses derives from the modern political profile of mineral production itself. States have found it necessary - largely for political reasons - to assume ever increasing control over all aspects of petroleum related activities on the continental shelf appertaining to them. The desire to take action to prevent oil pollution casualties, thus mitigating the serious damage

8. Claims to exercise fishery conservation jurisdiction beyond the limits of the territorial sea date from the early years of the century. The Icelandic claims, which increased from 12 nautical miles in 1960 to 200 nautical miles by 1976, virtually forced other states to follow suit as a protective measure, thereby laying the foundations for the regime of the Exclusive Economic Zone in international law.

9. The International Court of Justice has stated that "... the most fundamental of all rules relating to the continental shelf ... (is) that the rights of the coastal state (with respect to) the area of continental shelf that constitutes a natural prolongation of its territory into and under the sea exist ... by virtue of its sovereignty over the law, and as an extension of it ....", North Sea Continental Shelf Cases, [1969] I.C.J. Rep. 4 at 22.

10. The existing regime was a codification and development of the law as it had evolved by state practice from 1945-58, Convention on the Continental Shelf, April 29, 1958, in U.N. Doc. A/CONF 13/L.55 [hereinafter Continental Shelf Convention]. The law has undergone further development at the Third United Nations Conference on the Law of the Sea.

inflicted by such disasters, is another. These jurisdictional claims may affect the accepted tenets of traditional bodies of law, such as admiralty.[11]

In domestic or municipal law terms, the validity of claims to regulate mineral exploration and production on the continental shelf, or to take jurisdiction to prevent oil pollution in adjacent areas, is largely a matter for the legal system of the nation asserting those claims. Thus, generally speaking, if Canada, through the usual domestic law-making process, enacts such legislation, the courts will enforce those laws. However, whether other states will recognize the validity of such laws vis-a-vis their nationals is another matter. Validity is important where the legislation in question is deemed to apply to foreigners and nationals alike. Objections to a particular claim to jurisdiction will be unavailing where international law, reflecting the will of the international community, has sanctioned it. Accordingly, application of such laws to foreigners will be permissible at international law.

International law-making is very much an evolutionary process, frequently beginning with a single claim by one state, later followed by other states to the point that the claim becomes widely accepted, and may be codified and developed in a written convention.[12] A conference convened in 1958 under the auspices of the United Nations codified the law of the sea to a great extent in four Conventions on: the Territorial Sea and Contiguous Zone, the High Seas, the Living Resources of the High Seas,[13] and the Continental Shelf. This codification exercise was only a partial success, and a second conference in 1960 failed to

---

11. A number of conventions dealing with liability and compensation for oil pollution damage caused by ships and intervention on the high seas to prevent or mitigate a potential pollution disaster have been promulgated, and will be discussed infra. These conventions inherently recognize that an inherently transnational problem requires transnational remedial action, hence blurring to some extent the traditional ship - flag state relationship. Similar jurisdictional problems may arise regarding mobile drilling units, which may be registered under the maritime code of one state, but may be subject to the laws of another when drilling on foreign sectors of the continental shelf.
12. This process is well illustrated by state practice with respect to submarine mineral resources since 1945, which lead to the adoption of the Convention on the Continental Shelf in 1958.
13. Convention on the Living Resources of the High Seas, April 29, 1958, in U.N. Doc. A/CONF 13/L.54.

resolve some outstanding problems.[14] The Third United Nations Conference on the Law of the Sea (UNCLOS III), which has been meeting since 1974, concluded in December 1982 with the Final Act of the Conference and signature of the United Nations Convention on the Law of the Sea.[15] It is intended that this single treaty will, in time, replace the 1958 regime.

The validity of international agreements derives from consent as manifested by ratification (a process beyond signature signifying willingness to be bound). In Canada, ratification of a treaty is an executive act which does not require parliamentary consent. The Law of the Sea Convention will bind states party to it (i.e., those who have ratified it) twelve months following the deposition of the sixtieth instrument of ratification with the Secretary General of the United Nations. The Convention, though not authoritative in itself, can nevertheless be taken to represent the international consensus on the law of the sea for the purpose of this paper.

At this point, a somewhat paradoxical feature of the international law-making process should be noted. New principles of international law typically originate with a unilateral claim by one state, followed by broadly similar claims by others. But the principle that finally emerges and becomes accepted by the international community may deny the validity of these early claims which triggered the process. The origins of the doctrine of the continental shelf lie in a variety of differing claims to offshore jurisdiction. Extensive areas of adjacent hydrospace, usually expressed as claims to a 200 nautical mile territorial sea, were claimed by Central or South American countries, for example. Such claims would normally include sovereignty over the adjacent airspace.

---

14. See Derek W. Bowett, "The Second United Nations Conference on the Law of the Sea," 9 I.C.L.Q. 415 (1960).
15. United Nations Convention on the Law of the Sea, December 6, 1982, in U.N. Doc. A/CONF 62/122 [hereinafter Law of the Sea Convention], reprinted in XX1 Int'l Leg. Mat. 1261 (1982). The Final Act of the Conference appears in the same volume at 1245. The Convention was signed by Canada, but not by the United States of America and several western European countries, who remained dissatisfied with the deep seabed mining provisions. The Convention will enter into force twelve months following deposition of the sixtieth instrument of ratification (Article 308).

These claims continue to be lawful according to the domestic law of the nation in question and would be enforced by its courts. The fact, however, that the claim subsequently proves to be contrary to international law may well act as a strong political influence on states to modify their claims so as to conform with the new international principle. This has in fact happened with some claims to extensive areas of territorial sea, where the right of overflight by foreign aircraft is expressly upheld.

The international law of the sea has reached another important stage of development with the conclusion of UNCLOS III. The Law of the Sea Convention should, if accepted, set the seal of recognition on the coastal state's rights in the adjacent offshore for purposes which go well beyond the resource jurisdiction created by the regime of the continental shelf. At the same time, firm limits to that jurisdiction will be set.[16] This new regime is one of the reactions of the international legal order to the increasing demands placed upon it by offshore operations. The new regime will also attempt to ensure international supervision of submarine natural resources beyond the limits of national jurisdiction. Similarly, Canadian law and policy is evolving to meet new, nationally perceived needs. The Canada Oil and Gas Act,[17] the Arctic Waters Pollution Prevention Act,[18] and the federal Environmental Assessment and Review Process,[19] are all examples of this development.

---

16. Law of the Sea Convention, Part XI. Mere conclusion of the Treaty does not guarantee that coastal states will respect the limits set therein. This will be accomplished by self-enforcement and vigorous resistance by the International Seabed Authority and the United Nations to encroachments on marine or submarine areas which are part of the "common heritage of mankind". Failure to resist possible future claims may result, for example, in 200 nautical mile territorial seas, or claims to fisheries jurisdiction beyond 200 nautical miles. While the Convention may be taken to codify existing international customary law to a great extent, customary law may develop contrary to the intentions of the Convention.
17. Canada Oil and Gas Act, S.C. 1980-81-82-83, c.81.
18. Arctic Waters Pollution Prevention Act, R.S.C. 1970 (1st Supp.), c.2.
19. This process, discussed infra, was instituted by the decision of the Federal Cabinet on December 20, 1973, and is outlined in the Revised Guide to the Federal Environmental Assessment Review Process, Ottawa, Federal Environmental Assessment Review Office, 1979.

CANADIAN BACKGROUND

Canada, like all states, wishes her domestic legal system to remain in harmony with international law. The entry into force of the Law of the Sea Convention may therefore have an impact on domestic law already in place, especially if ratified by Canada. In turn, this process may have an impact on the uncertainty which attaches to some of Canada's jurisdictional claims in the Arctic. One of the purposes of this paper is to examine in outline the legal issues arising from these uncertainties in the specific context of Canada's jurisdiction over oil and gas operations in Arctic waters.

Exploitation of the petroleum resources of offshore areas in the Arctic, although of interest to relatively few states, nonetheless must take place within the legal framework for offshore development fixed by international law. In the Arctic, as with almost any other part of the world's oceans, other users of the sea may be affected by the operations of the petroleum industry. It is up to the legal system, both international and domestic, to attempt to balance the various rights and claims of these sea users; in the domestic context the political impetus imparted to a particular activity is likely to result in a completely new balance being struck.

The position in the Arctic is further complicated by the fact that this area is not a busy seaway like the English Channel, the Straits of Gibraltar, or the Panama Canal. The waters of the Canadian Arctic are rarely used by shipping. However, they are used, in a unique fashion, by the Inuit, for whom access to the living resources is an integral part of their way of life, not susceptible to measure according to the economic criteria established by our legal system. On the other hand, the "way of life" argument also applies to east coast fishermen.

The oil industry was first encouraged to consider the petroleum production potential of the Canadian Arctic by the Conservative government of 1958-1962. As part of his "vision of the north", Prime Minister Diefenbaker was anxious to further exploration for oil and gas in the Arctic. Petroleum operations, onshore and offshore, had already been in progress for some decades. The first onshore discovery, Norman Wells, was made in 1921, one year after exploratory drilling in the Northwest Territories commenced. Offshore exploration started in 1961.

OFFSHORE ACTIVITY FROM 1973 TO THE PRESENT

The discovery of the Adgo well by Imperial Oil in 1974 was the first in the Canadian Arctic offshore; drilling had started at Immerk the previous year. Exploration programs had been in progress since the late 1950s, but early drilling, not surprisingly, was landbased. Drilling from drillships commenced in 1976, and operations from the first artificial island began in 1981.

Canada Oil and Gas Lands Administration (COGLA) statistics[20] indicate that five offshore wells are currently being drilled in Arctic waters, four from drillships and one from a caisson-retained island, by the three main operators in that area, Dome Petroleum Ltd., Esso Resources Canada Ltd., and Gulf Canada Resources Inc.

CANADIAN LEGISLATION FROM 1960 TO THE PRESENT

The Canada Oil and Gas Regulations,[21] issued in 1961, reflected the wishes of the industry to a very great extent, combining tax and royalty provisions with generous tax write-offs as an incentive to exploration, and far from onerous work obligations. The regulations were issued pursuant to the Public Lands Grants Act[22] and the Territorial Lands Act.[23]

Encouragement of this kind was clearly no substitute for commercial discoveries, but the attitude of the industry to the Canadian north changed drastically in 1970 with the discovery of a massive oil and gas field near Prudhoe Bay, Alaska. Exploration in the Canadian Beaufort Sea intensified following this find; the first of many artificial islands, Immerk, was constructed in 1972. Shortly after this it became clear that the Prudhoe Bay discovery was perhaps exceptional rather than the rule - one large reservoir, rather than a series of much smaller reservoirs.

---

20. Reproduced in the Administration's Submission to the Special Committee of the Senate on the Northern Pipeline, September 14, 1982 at 24 to 28. The Submission also contains a comprehensive outline of the structure, responsibilities, and activities of COGLA itself.
21. Canada Oil and Gas Regulations, SOR/61-253 as am.
22. Public Lands Grants Act, R.S.C. 1970, c.P-29.
23. Territorial Lands Act, R.S.C. 1970, c.T-6.

The Canada Oil and Gas Regulations were revised in 1977;[24] this revision was intended as a temporary measure prior to the enactment of a new oil and gas licensing regime for the Canada Lands.[25] A Bill[26] to implement the new system was introduced into Parliament in the same year, but lapsed when Parliament was prorogued in 1979. Bill C-48, introduced into Parliament in 1980, received royal assent as the Canada Oil and Gas Act in March 1982.[27] The Act implements a new licensing regime for oil and gas exploration and production on the Canada Lands, and effected amendments to the Oil and Gas Production and Conservation Act vis-a-vis liability for oil spills.

The Act was promulgated pursuant to the National Energy Program (discussed briefly in the section "Control of Resource Exploration and Production") which seeks, inter alia, to secure the exploitation of Canadian resources by Canadian companies for the benefit of Canadians. The Act contains strict requirements for Canadian ownership and Canadian participation in ventures on Canada Lands, and entities which do not match up to these requirements will not be eligible for incentive payments; entities may lose their right to hold licences under the new system if the Canadian ownership rate falls below a certain point.[28] It is clear, therefore, that the federal government looks to Canadian companies to justify its policy decisions with regard to Canada Lands by intensifying the search for the petroleum resources of the Canadian Arctic.

International law can only provide the means for the settlement of problems arising out of the co-relative rights and duties of states with respect to offshore petroleum exploration and development. But implementation of these rights is a

---

24. Canada Oil and Gas Land Regulations, C.R.C. 1978, c.1518.
25. The "Canada Lands" comprise the Yukon Territory, the Northwest Territories, and the seabed and subsoil of offshore areas over which Canada exercises mineral resources jurisdiction. See the section "Control of Resource Exploration and Development".
26. Bill C-20, 3rd Sess., 30th Parliament, 26 Elizabeth II, 1977.
27. The Canada Oil and Gas Act, S.C. 1980-81-82-83, c.81. The new licensing regime is discussed in Harrison, Rowland J. and Townsend Gault, Ian, Bill C-48 - A Framework for Comparative Analysis, Calgary, Canadian Institute of Resources Law, The University of Calgary, 1981, and in Townsend Gault, Ian, Petroleum Operations on the Canadian Continental Margin - The Legal Issues in a Modern Perspective, Calgary, Canadian Institute of Resources Law, The University of Calgary, 1983 at 79 to 94.
28. See Canada Oil and Gas Act, S.C. 1980-81-82-83, c.81, s.20 et seq.

matter for each state to decide for itself. In particular, it is up to states to adjudicate among the use of particular sea areas. Thus, it will be up to Canadian authorities, provincial and federal, to decide ultimately the extent to which offshore oil development of east and west coasts interferes with established practices of fishing and navigation in these areas. Similarly, the same sort of balancing act must be performed in Arctic waters.

Now that the international legal regime and domestic licencing policy for offshore areas have moved into a new phase almost simultaneously, concern about the impact of offshore development on the social, economic, and environmental framework of the Arctic is resurfacing. The residents of the two territories are becoming increasingly vocal, demanding not only the right to be heard, but also to participate, when important decisions concerning the pace and location of development are being made.

There is another cause for concern. Canada is commencing offshore development in a state of what can only be described as juridical unpreparedness. Perhaps because of the federal-provincial dispute concerning jurisdiction over offshore resources, much of the legal debate regarding offshore operations has tended to concentrate on the vexed question of the right to control development and reap the benefits accruing therefrom.

This rather narrow focus tends to ignore the fact that just as offshore development is being imposed on the existing uses of offshore areas, it is also being imposed on the existing legal system. Special rules may be implemented for many aspects of offshore development, but these special rules are not necessarily comprehensive. The existing legal structure remains as before, a structure which allocates rights and involves responsibilities between the two levels of government. In the Arctic, the two levels of government become one for most purposes; the totality of traditionally accepted areas of government responsibility remains. A whole new set of problems arise when the existing legal framework comes under pressures and demands it is unable to meet, and has not been designed to meet. Legislative neglect may be matched by judicial initiative which sets precedents for the future. If these precedents are unfortunate, the impact on the development of Canadian law as a whole is similarly deleterious.

SCOPE OF THE PAPER

In order to take stock of the legal position in the north, this paper will present various aspects of Canada's claims, and the equivalent international legal regime, in an itemized form. In this, the first part of the paper, I have attempted to set the scene in international and Canadian terms, for the discussion to follow. The second part presents an overview of the principle jurisdictional claims asserted by Canada in the Arctic, summarizing these claims from an essentially domestic, Canadian, legal perspective. The third and main part of the paper sets Canada's claims in an international law context and articulates the issues involved in testing their validity from the international perspective. In the last part some conclusions on the present state of the law are advanced.

AN OVERVIEW OF CANADA'S PRINCIPAL JURISDICTIONAL CLAIMS IN ARCTIC WATERS

In the context of oil and gas operations in Canada's northern offshore areas, claims with respect to three functions are of particular relevance. These functions are: exploration for and production of the resource itself, environmental protection, and transportation by pipeline and tanker. The relevant provisions of the Canada Shipping Act with respect to the coasting trade are also noted in this context. In this part, Canada's claims with respect to each of these functions are summarized and the foundation of these claims in the domestic legal system is identified. However, other related claims, with respect to the territorial sea and fishing zones, and with respect to jurisdiction over the Arctic islands, must be examined first in order to explain the development of this country's claims in the offshore. Environmental protection raises additional jurisdictional issues and problems vis-a-vis vessels, pipelines, and installations, and these are also examined. Finally, problems arising with respect to court jurisdiction in offshore areas, and with offshore installations and facilities, are discussed.

CLAIMS TO THE ARCTIC ISLANDS

International law recognizes claims to territory based on a number of criteria such as discovery, occupation, and conquest. What constitutes "occupation" may vary from case to case; a state may "occupy" an uninhabited and uninhabitable island by the mere formality of taking possession. The claim is further enhanced by inclusion of the island on official maps, showing it to be part of the state's territory, or, for example, the maintenance of meteorological stations or navigational aids on the island.

Canada's claim to her Arctic archipelago is beyond question. There is a wealth of evidence demonstrating continuous occupation, and the claim has not been challenged by any other state. There are some disputes over individual islands: both Canada and Greenland claim Hans Island in the Horneby Channel, which is almost equidistant from Ellesmere Island and Greenland. Possible consequences of this dispute with respect to oil and gas activities are outlined below.

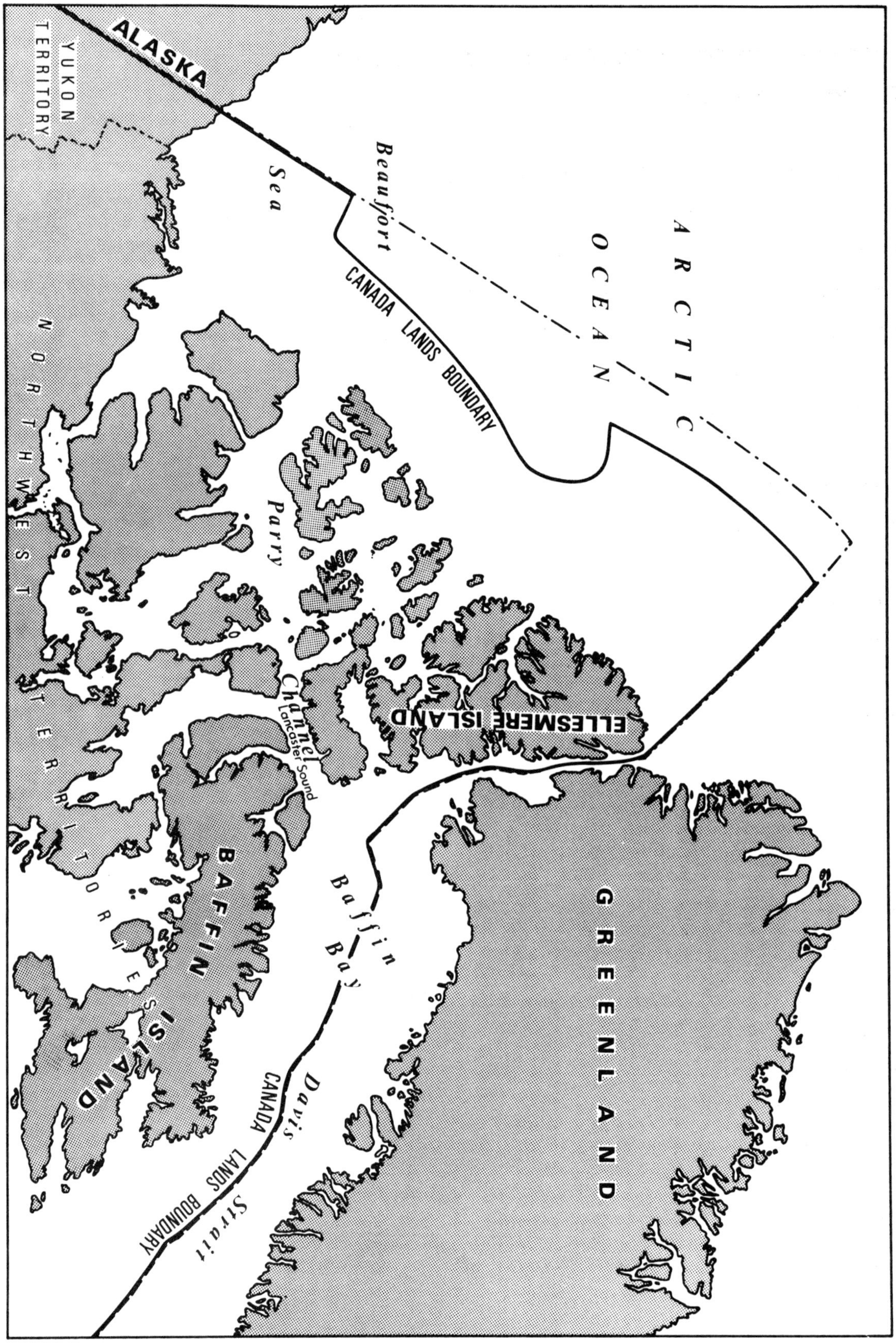

Figure 1. The Arctic Offshore

THE TERRITORIAL SEA

Canada claims a territorial sea of twelve nautical miles breadth[29] over which she exercises sovereignty subject only to the right of innocent passage exercisable by foreign vessels. The breadth of the territorial sea is not measured from the coast as such, but from baselines, which usually correspond to the low-water mark.[30] Domestic law *may* apply in the territorial sea as though the area were part of the landmass of the nation, but the extent to which it does is for each state to determine. In Canada, legislation is deemed not to apply seaward of the baseline unless the contrary is clearly indicated.[31] Canada has, at least since 1958, regarded the waters of the Arctic archipelago as internal waters,[32] over which she has complete sovereignty as if they were inland waters. Since baselines have not been drawn around the islands as a group, this claim must be based on historic title, i.e., a claim that the waters have historically been regarded as Canadian.[33] Other Canadian claims are bound up with this, such as the jurisdictional zone

---

29. Territorial Sea and Fishing Zones Act, R.S.C. 1970, c.T-7.
30. Territorial Sea Convention, Arts. 3, 4.
31. Section 433 of the Criminal Code, R.S.C. 1970, c.C-34, clearly extends Canadian criminal law to the territorial sea. A number of other Canadian statutes also apply, either expressly or by implication, to the territorial sea and are listed in A.L.C. de Mestral, "The Law Applicable to the Canadian East-coast Offshore", XXI Alta. L. Rev. 63 (1983).
32. The Minister of Northern Affairs informed the House of Commons in 1958 that "(t)he area to the North of Canada, including the islands and the waters between the islands, and the areas beyond, are looked upon as our own ... this is national terrain." This statement was quoted by Prime Minister Trudeau in a statement to the House on May 15, 1969, as representing the position of his government also - in: Canada. House of Commons. Debates, Vol. VIII at 8720 (1969). For discussion of the implications of this view with respect to the rights claimed by Canada seaward of the archipelago, see text at note 165, *infra*.
33. It is suggested that failure to clarify the status of the archipelagic waters is likely to result in increased juridical confusion. A recent example of this confusion is evidenced by *dicta* of Bourassa, J., in *R.* v. *Panarctic Oils Ltd.*, (January 14, 1983, N.W.T.S.C., unreported) where His Lordship appeared to regard the waters of the Arctic archipelago as Canadian territorial waters, "...using a straight baseline approach to encircle the Archipelago referred to in the *Anglo-Norwegian Fish Case* (sic) ... (and because) ... Canada's jurisdiction over the waters is accepted as part of international law." The point is, of course, that straight baselines have not been drawn, and if they were then the waters of the archipelago would be internal waters. Neither party argued the question of jurisdiction in the above case.

promulgated by the Arctic Waters Pollution Prevention Act, and the sector theory. If this claim can be sustained, a number of jurisdictional problems noted below will be more readily resolved.

CONTROL OF RESOURCE EXPLORATION AND PRODUCTION

LIVING RESOURCES

Canada exercises ownership over internal waters and the territorial sea, the seabed and subsoil, and the superjacent airspace. Canada may assert ownership over the living resources of these marine areas also. However, Canadian law holds that fish, like wild animals, are *res nullius*, the property of no one in their natural state, and must be reduced to possession. Title flows from capture, as with oil and gas itself.[34] However, as with oil and gas, Canadian law makes provision for the circumstances in which living resources such as fish may be exploited, and has done so through regulations made pursuant to the Territorial Sea and Fishing Zones Act under which a two hundred nautical mile exclusive fishing zone has been established.[35] Such zones are frequently claimed by coastal states to enforce conservation measures, or to protect local industry. Fishing zones will be subsumed into the regime of the Exclusive Economic Zone when the Law of the Sea Convention enters into force.[36] The fishing zone as a concept, however, has little relevance in the context of oil and gas operations.

---

34. See the leading case of *Berkheiser* v. *Berkheiser*, [1957] S.C.R. 387, where the Supreme Court of Canada rejected the argument that a petroleum lease was a sale of minerals, but rather granted a *profit a prendre* or right to take the minerals; ownership would then flow from possession.
35. The zone in northern waters was created by the Fishing Zones of Canada (Zone 6) Order, C.R.C. 1978, c.1549.
36. The rights of coastal states in the zone include the management, conservation, and exploitation of living resources - Convention on the Law of the Sea, Part V, Art. 56(1)(a).

NONLIVING RESOURCES

The National Energy Program

The National Energy Program (NEP) was published in October 1980 by the Deparment of Energy, Mines and Resources. A major policy statement covering exploration and development of nonrenewable resources on Canada Lands, the goals of the Program are the stimulation of such activities by entities owned or controlled by Canadians, including Crown corporations, in order to achieve the goal of energy self-sufficiency.

One of the basic tenants in the government's argument justifying radical changes to the status quo was the claim that existing licence conditions were too "lenient", and were incapable of stimulating activity to the pace which the government now required. The solution would therefore be to implement new legislation amending the licensing regime, apply it to existing interest holders, increase state participation through a Crown corporation, provide incentives, and increase the extent of discretionary government control.

Three aspects of the implementation of the NEP relevant in the present context will now be examined: the Canada Oil and Gas Act, the Canada Oil and Gas Lands Administration, and exploration agreements in Arctic Waters. The relationship between the NEP and other policy goals in the Arctic will be discussed in the course of the text.

The Canada Oil and Gas Act

The seaward extent of Canada's offshore claim is now set by the Canada Oil and Gas Act, which supercedes the Canada Oil and Gas Land Regulations promulgated under the Territorial Lands Act[37] and the Public Lands Grants Act.[38] Pursuant to the regulations, Canada has issued exploration permits for areas in excess of two hundred nautical miles offshore. The federal government is currently considering

37. Territorial Lands Act, R.S.C. 1970, c.T-6.
38. Public Lands Grants Act, R.S.C. 1970, c.P-29.

proposals from candidates for further exploration agreements to be granted pursuant to the Canada Oil and Gas Act, while renegotiating existing permits and licences.[39]

The Act implements a new licensing regime for the "Canada Lands" which are defined to include: the Yukon Territory, the Northwest Territories, and

> ... those submarine areas, not within a province, adjacent to the coast of Canada and extending throughout the natural prolongation of the land territory of Canada to the outer edge of the continental margin or to a distance of two hundred nautical miles from the baselines from which the breadth of the territorial sea of Canada is measured, whichever is greater.[40]

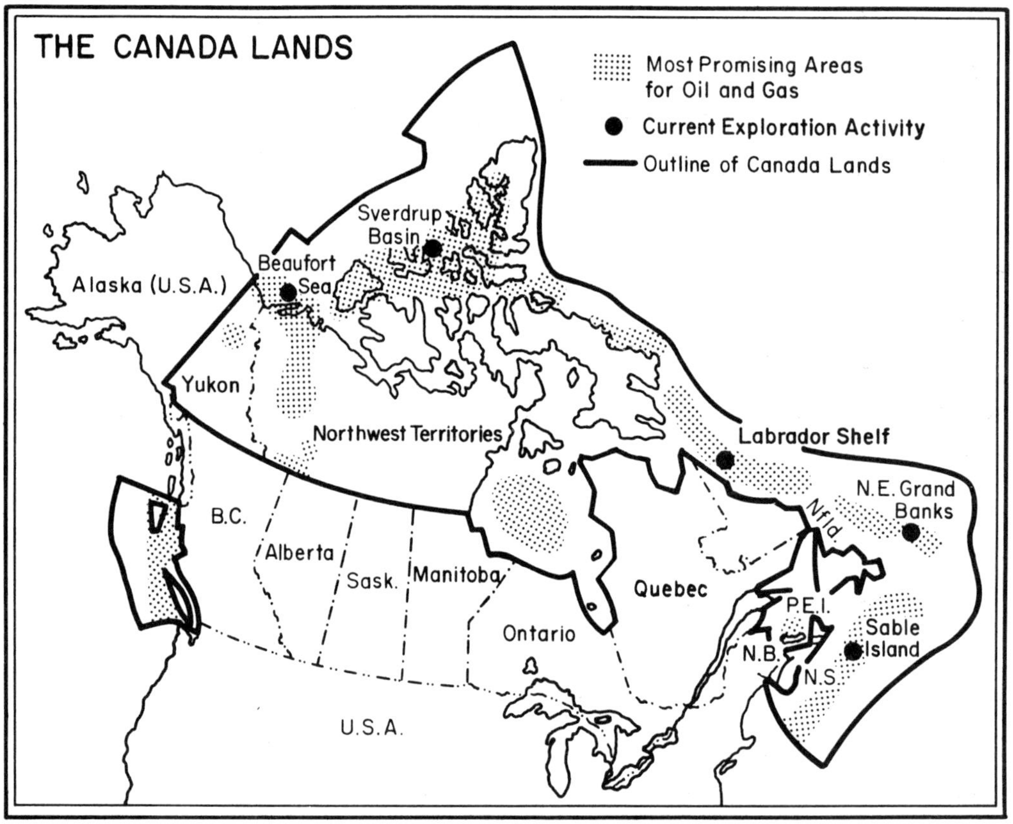

Figure 2. The Canada Lands

---

39. Canada Oil and Gas Act, S.C. 1980-81-82-83, c.81, s.63 <u>et seq</u>.
40. <u>Id</u>., s.2(1).

Canada therefore claims exclusive control of mineral resources to a minimum distance of two hundred nautical miles and beyond that to the outer edge of the continental margin where that margin so extends. Hence, questions concerning jurisdiction in marine areas and the Arctic islands must be settled in order to determine those fixed offshore zones measured from "land".

The new licensing system was implemented to meet contemporary needs as laid out in the National Energy Program.[41] But existing interests were to be surrendered or brought under the new regime.[42] This "fundamental restructuring" would result in "a modernized regulatory basis for the management of (the Canada Lands)".[43]

The basic structure of the previous licensing regime is retained. Nonexclusive exploration rights may be granted under the Oil and Gas Production and Conservation Act. Exclusive exploration rights and the right to produce oil and gas if found in commercial quantities are granted under the Canada Oil and Gas Act by instruments which are essentially contracts, although the minister's discretionary powers are considerable.[44]

No activity on Canada Lands may be pursued without an operating licence,[45] which is the instrument used to regulate prospecting (i.e., exploratory activities short of drilling). The terms and conditions of a licence are agreed to between the minister and the applicant, but the Act provides for expiry on the March 31st following issue, but it may be renewed for further periods of up to one year.[46]

---

41. Supra, note 1.
42. Canada Oil and Gas Act, S.C. 1980-81-82-83, c.81, s. 61.
43. National Energy Program, supra, note 1 at 45.
44. Apart from the grant of the interest itself, the minister determines the commercial viability of a field, he must approve the development program for a field, and he is empowered to order interest holders to drill further wells and deliver petroleum produced therefrom to any part of Canada at a price set by him - Canada Oil and Gas Act, S.C. 1980-81-82-83, c.81, ss.44, 45, and 48. Authorization in writing is required before any proposed activity is commenced - Oil and Gas Production and Conservation Act, R.S.C. 1970, c.O-4, s.3.1.
45. Oil and Gas Production and Conservation Act, R.S.C. 1970, c.O-4, s.3.1.
46. Id.

Exploration agreements are negotiated between applicants and the minister, but the Canada Oil and Gas Act sets out certain terms and conditions concerning the qualifications of applicants[47] and the system of allocation.[48] The agreement invests the holder with the exclusive right to enter onto the licensed area to explore for and produce oil and gas therein.[49] A work program will normally be required, and provisions for "Canada benefits" specified.[50] Again, the specific details of the agreement, including the work program (which will obligate the applicant to drill a certain number of wells to specified depths), are a matter for negotiation between the parties.[51] In the event of a commercial discovery of petroleum, the applicant has the right to apply for a production licence, which will be issued on approval of a development plan for the discovery. It is also highly probable that in the event of the possible production of oil and gas, the federal Environmental Assessment Review Process (discussed below) will be initiated.

A production licence invests the holder with the exclusive right to produce oil and gas.[52] The licence period is ten years, renewable for further ten-year periods while the field remains commercially viable.[53] A basic royalty of 10% is payable on petroleum produced[54] together with a progressive incremental royalty of 40% of the net profit accruing to the licensee.[55] Assignment or subdivision of interests is subject to ministerial approval[56] and registration.[57]

---

47. Canadian corporations as defined in the Act.
48. The Act assumes agreements will be concluded following calls for public tender (s.11), but the minister may enter into agreements at any time in the public interest (s.14). The minister is not bound to select any proposals submitted following a call (s.14).
49. Canada Oil and Gas Act, S.C. 1980-81-82-83, c.81, s.9.
50. Id., s.10(2). Such benefits would include the employment of Canadians, and the use of Canadian goods and services. Exploration agreements concluded in Arctic areas have included native training schemes, for example, as part of the Canada benefits package.
51. Outlines of the seventy exploration agreements concluded as of January 31, 1983, indicate that minimum drilling commitments total some 48 wells (see COGLA statistics appended to the announcement of an outline agreement concluded with Gulf Canada Resources Inc., January 20, 1983).
52. Canada Oil and Gas Act, S.C. 1980-81-82-83, c.81, s.17. Only permanent Canadian residents or Canadian corporations with a minimum Canadian ownership rate of 50% may hold production licences, id., ss.18, 19.
53. Id., s.25.
54. Id., s.40.
55. Id., s.41.
56. Id., s.52.
57. Id., s.53.

The Canada Oil and Gas Act mades no reference to the limits of Canadian jurisdiction constituted by maritime boundaries with other states. When concluded, these agreements should constitute additional limits to this country's offshore jurisdiction.[58]

As long as the Yukon Territory and the Northwest Territories remain subject to the jurisdiction of the federal government, jurisdiction over the resources of the Canadian Arctic waters will not be in dispute as it is off the east and west coasts. However, there is one jurisdictional issue that may be of some importance in the future. It would seem from the definition of "Canada Lands" above, that Ottawa claims jurisdiction over the petroleum resources of three distinct areas: the Canadian continental margin south of 60 degrees, the Yukon and Northwest Territories, and the continental margin of the two territories. The first area comes within the jurisdiction of the federal Department of Energy, Mines and Resources; the other two under the Department of Indian Affairs and Northern Development (DIAND).

The Canada Oil and Gas Lands Administration

The Canada Oil and Gas Act marks an intensification of government control over all aspects of offshore oil development. That increased supervision and control is being exercised through a new government agency, the Canada Oil and Gas Lands Administration (COGLA). It is COGLA's task to administer the Act, and to supervise many aspects of hydrocarbon development on all Canada Lands. To this

58. One agreement with Denmark (Greenland) is already in place, Agreement on the Delimitation of the Continental Shelf between Greenland and Canada, Ottawa, December 17, 1973, in XIII Int'l Leg. Mat. 506 (1974); Lay, S. Houston, comp., New Directions in the Law of the Sea, New York, Oceana Publications, Inc., Vol. I - , 1973- [hereinafter New Directions], Vol. IV at 105. The treaty entered into force on March 13, 1974. See section "Offshore Boundaries".

end, it is responsible jointly to the Ministers of Energy, Mines and Resources, and Indian Affairs and Northern Development.[59]

COGLA is headed by an Administrator, who is responsible for the performance of a variety of management functions under the Canada Oil and Gas Act, and who has also been designated as Chief Conservation Officer for the purposes of the Oil and Gas Production and Conservation Act.[60] The Administration has six branches: Land Management, Engineering and Control, Resource Evaluation, Environmental Protection, Canada Benefits, and Interdepartmental Relations. Land Management is responsible for the negotiation and issuance of exploration agreements and production licences, registration of interests, approval of interest transfers, and collecting and recording the government "take". Engineering supervises and monitors drilling, production, and conservation, and evaluates production systems proposed for use on Canada Lands. Resource Evaluation has a broad responsibility for evaluating hydrocarbon prospects, monitoring information submitted by operators, and the identification of natural hazards which might influence the choice of exploration, production, or transportation systems.

The Canada Benefits branch is responsible for negotiating terms in exploration agreements pertaining to use of Canadian (or local) goods, services, and labour. The Environmental Protection branch is empowered to negotiate special terms with respect to the protection of particular areas and the conduct of activities to be carried on therein. These two areas of responsibility are of particular interest with respect to operations in the Canadian Arctic.

There are those who consider that this arrangement may mean that oil and gas development in the territories and in Arctic waters will be determined by nationally perceived requirements to such an extent that previously identified goals of local development, and the potentially harmful economic, social, and

---

59. COGLA was established by the Memorandum of Understanding between the Ministers of Energy, Mines and Resources, and Indian Affairs and Northern Development of August 21, 1981, and a Letter of Agreement between the Deputy Ministers of the two departments concluded the same day. The Administration discharges the responsibilities of the respective Ministers under the Canada Oil and Gas Act and the Oil and Gas Production and Conservation Act.
60. The Administrator has a number of other functions; he is also Chairman *ex officio* of the Canada-Nova Scotia Offshore Resources Board, for example.

environmental consequences to the north and its people, will go unrecognized until the problem reaches serious proportions. The Arctic has, according to some writers, seen a similar jurisdictional struggle between the Department of the Environment and the Department of Indian and Northern Affairs in the late 1960s and early 1970s.[61] The struggle arose when the former was attempting to influence the course of petroleum development in the two territories as it affected the environment. That development, however, came under the jurisdiction of the latter, which appeared anxious to prevent such concerns from delaying the exploration for hydrocarbons. The Department of the Environment was adjudged to have been largely unsuccessful; there are fears that history may be about to repeat itself.

It remains to be seen how well DIAND can control that part of COGLA's activities which pose or which might pose these sorts of threats to the Arctic and its inhabitants. This is very much a political or public policy issue, not a legal one.

Exploration Agreements in Arctic Waters

Six exploration agreements have been concluded with a consortium led by Esso Resources Canada Ltd. covering acreage in the Mackenzie Delta and in the Beaufort Sea.[62] The length of term of the agreements, three each for onshore and offshore areas, is five years commencing June 1, 1982. The total area covered will be some six million acres.

The work commitments for the offshore agreements specify a minimum exploration effort. Seismic work will generally be required, but the outline agreements set out the drilling requirements in more detail. Esso is required to drill at least five wells, one in each area, to depths sufficient to permit evaluation of hydrocarbon production potential. As the elements of the work program are

---

61. See Pimlott, Douglas H., Brown, D., and Sam, K.P., Oil Under the Ice, Ottawa, Canadian Arctic Resources Committee, 1976 at 17 *et seq.*
62. These details have been abstracted from a DIAND press release of May 13, 1982.

completed, Esso's entitlement to retain the interest in any discovery will increase, subject only to the acreage relinquishment provisions.[63]

The outline agreement specifies formulae for the identification of acreage to be relinquished, up to a maximum of 50% by the fourth anniversary of the commencement of term (unless Esso wishes to relinquish further acreage voluntarily). The rationale behind the formulae is the need to strike a balance between Esso's wish to retain a manageable area including the best production prospects and the Crown's desire to have a relinquished area which is capable of attracting future exploration agreement applicants. Environmental programs will also be required of the agreement holders; details of these programs have not as yet been released.

The single agreement with Gulf Canada Resources Inc., announced in January 1983, covers 0.6 million hectares of the Beaufort Sea northwest of Tuktoyaktuk, and prescribes a minimum of five wells to be drilled during the five year term.

ENVIRONMENTAL PROTECTION

Many aspects of environmental protection are considered an integral part of the regulatory regime governing resource exploration and production. As such, environmental protection in this context generally coalesces with resource exploration and production so far as the extent of Canada's jurisdiction is concerned. In the north, however, Canada has made a unique claim to exercise jurisdiction with respect to environmental protection which is highly relevant to resource exploration and production - the Arctic Waters Pollution Prevention Act, which is discussed in the next section, "Transportation By Tanker".

The federal Environmental Assessment Review Process (EARP), instituted by Cabinet directive in 1973, provides for the screening of proposed activities which may impact on the environment, to be undertaken at the discretion of the Minister

63. The Canada Oil and Gas Act makes no provision for acreage relinquishment, but such provisions were incorporated into the previous regime, and are common in other jurisdictions; see Harrison and Townsend Gault, _supra_, note 27.

of the Environment. Clearly, oil and gas operations, onshore and offshore, and transportation of petroleum by land, sea, or pipeline, may come within this category.

The EARP process makes provision for a review of the environmental implications of a proposed project by a Panel nominated by the minister. Federal government departments are required to consider the environmental consequences of projects falling within their sphere of responsibility. If a project cannot be allowed to proceed after "screening" (because, for example, the environmental consequences are not fully known), an Initial Environmental Evaluation takes place. If the project cannot then be cleared, investigation by a Panel is instigated.

This procedure has been followed with respect to proposals to explore for oil in Lancaster Sound[64] and Davis Strait,[65] and also with respect to the Arctic Pilot Project.[66] The process is currently underway with respect to the proposed production of oil and gas from the Hibernia oilfield and in the Canadian Beaufort Sea.[67] The assessment process for this proposal was initiated by the Department of Indian Affairs and Northern Development, and the project was referred for review by a Panel on July 22, 1980.

---

64. Lancaster Sound Environmental Assessment Panel. Lancaster Sound Drilling: report, Federal Environmental Assessment and Review Process 7, Ottawa, Federal Environmental Assessment Review Office, 1979. Some of the options for hydrocarbon exploration, production, and transportation are discussed in: Working Group on the Lancaster Sound Region Study. Green Paper, The Lancaster Sound Region: 1980-2000: perspectives and issues on resource use, Ottawa, Indian and Northern Affairs, 1980.
65. Eastern Arctic Offshore Drilling Environmental Assessment Panel (South Davis Strait). East Arctic Offshore Drilling - South Davis Strait Project: report, Federal Environmental Assessment and Review Process 6, Ottawa, Federal Environmental Assessment Review Office, 1978. The North Davis Strait Project is not proceeding for the time being whilst the Department of Indian Affairs and Northern Development considers not only the Panel report on this project but also that on Lancaster Sound drilling.
66. Arctic Pilot Project Environmental Assessment Panel. Arctic Pilot Project (Northern Component): report, Federal Environmental Assessment and Review Process 14, Ottawa, Federal Environmental Assessment Review Office, 1980.
67. Beaufort Sea Environmental Assessment Panel. Beaufort Sea Hydrocarbon Production Proposal: interim report, Federal Environmental Assessment and Review Process 19, Ottawa, Federal Environmental Assessment Review Office, 1982.

Guidelines for an Environmental Impact Statement, issued by the Panel, were publically reviewed in late 1981. The proponents, Dome Petroleum Ltd., Esso Resources Canada Ltd., and Gulf Canada Resources Inc., responded to these guidelines in the Environmental Impact Statement (EIS) itself which was submitted in late 1982.[68]

The Panel's task was to review the Environmental Impact Statement submitted by the proponents, obtain public reaction thereto, and advise the minister of the acceptability, or otherwise, of the environmental effects of the project. The Panel's views are recommendatory only; the Minister of the Environment, and the minister of the initiating department (in this case Indian Affairs and Northern Development) decide on any action to be taken.

In the event, the Beaufort Sea Panel, while commending the proponents on their work in preparing the EIS, issued a Deficiency Statement which identified four major deficiencies therein.[69] These deficiencies are in the categories of socio-economic effect assessment (i.e., onshore impact on communities, and the monitoring and mitigation thereof), environmental effects assessment (i.e., the risk element and the environmental effects of the proposed development), issues arising from the possibility of oil spills, and summaries of the environmental impact in the three zones: Beaufort Sea - Mackenzie Delta, Mackenzie Valley, and the Northwest Passage. The Panel has requested the proponents to respond to the Deficiency Statement, after which public hearings can commence.

Implementation of a Panel's recommendations are effected by the appropriate ministers and the proponents. Canadian corporations may be bound by a ministerial decision in a variety of ways; a decision may, for example, be incorporated into the development plan for an oil or gas discovery. But jurisdictional problems may arise where these directions purport to apply, for example, to foreign ships outside Canadian territorial waters, or foreign vessels

68. The Environmental Impact Statement will be released in a number of volumes, some of which have already appeared; the original version of this paper was included in one such volume.
69. Beaufort Sea Environmental Assessment Panel. A Statement of Deficiencies on the Environmental Impact Statement for Hydrocarbon Development in the Beaufort Sea - MacKenzie Delta Region, Ottawa, March 1983.

availing themselves of the right of innocent passage through the Canadian territorial sea. However, the regime of the coasting trade (discussed in section "The Coasting Trade") might operate so as to prevent the use of foreign vessels for supplying offshore facilities, or for transportation of petroleum produced.

TRANSPORTATION OF OIL AND GAS

TRANSPORTATION BY PIPELINE

The construction and operation of pipelines in Canada is regulated by the National Energy Board Act.[70] The Board clearly has jurisdiction over interprovincial or international pipelines onshore, but it is not immediately clear whether there is a basis for taking jurisdiction offshore. Since there is no indication to the contrary, the Act must be presumed not to apply beyond the low-water mark or other baselines (i.e., beyond internal waters). The Act may, however, apply in the waters of the Canadian Arctic archipelago if those waters are indeed internal waters as claimed by Canada. It is doubtful whether the Board would have jurisdiction over pipelines on the Canadian shelf laid by other states or their nationals.

The federal government has introduced guidelines for the construction of pipelines in the north.[71] The legal status and jurisdictional extent of these guidelines is ambiguous - do they apply on the continental shelf, within territorial waters, or only to inland waters and onshore? This is another case where clear legislative provision is perhaps required.

Recent amendments to the National Energy Board Act[72] make provision for the Board to control the movement of petroleum from the offshore either to the mainland of Canada or elsewhere. The amendments are discussed in the next section, "Transportation By Tanker". No pipelines have as yet been constructed

---

70. National Energy Board Act, R.S.C. 1970, c.N-6.
71. Expanded Guidelines for Northern Pipelines, tabled in the House of Commons by the Minister of Indian and Northern Affairs, June 28, 1972.
72. National Energy Board Act, R.S.C. 1970, c.N-6; National Energy Board Amendment Act (No. 3), S.C. 1980-81-82-83, c.116.

for the transportation of petroleum on the Canadian continental margin. The debate on modes of transportation is still in progress, with options being mooted or promoted by the parties concerned.[73]

TRANSPORTATION BY TANKER

The legal regime applicable to Canadian ships, and to some extent, all shipping in Canadian territorial waters, is contained in the Canada Shipping Act.[74] Canada has, however, promulgated a special regime for Arctic waters through the Arctic Waters Pollution Prevention Act.

The Arctic Waters Pollution Prevention Act

This measure is primarily concerned with discouraging the unauthorized deposit of waste materials in Arctic waters, but also enables regulations dealing with tanker design and operational standards to be promulgated. The definition of "waste" includes any substance that would detrimentally alter, degrade, or contribute to the detrimental alteration or degradation of the water it is added to.[75]

The Act applies to Arctic waters which are those "adjacent to the mainland and islands of the Canadian Arctic within the area enclosed by the sixtieth parallel of north latitude, the one hundred and forty-first meridian of longitude and a line measured seaward from the nearest Canadian land a distance of one hundred nautical miles."[76] The Act does not apply to inland waters, and the extent of

73. *The Lancaster Sound Region: 1980-2000*, *supra*, note 64.
74. Canada Shipping Act, R.S.C. 1970, c.S-9, Part XVI (Collisions) and Part XX (Pollution).
75. Arctic Waters Pollution Prevention Act, R.S.C. 1970 (1st Supp.), c.2, s.1. See Pharand, Donat, "Oil Pollution Control in the Canadian Arctic", 7 Texas Int'l L.J. 45 (1971); the same author's *The Law of the Sea of the Arctic, with special reference to Canada*, Ottawa, University of Ottawa Press, 1973, Part VI; and J.A. Beasley, "The Arctic Waters Pollution Prevention Act: Canada's Perspective", 1 Syracuse J. Int'l L. & Com. 226 (1973).
76. Arctic Waters Pollution Prevention Act, R.S.C. 1970 (1st Supp.), c.2, s.2.

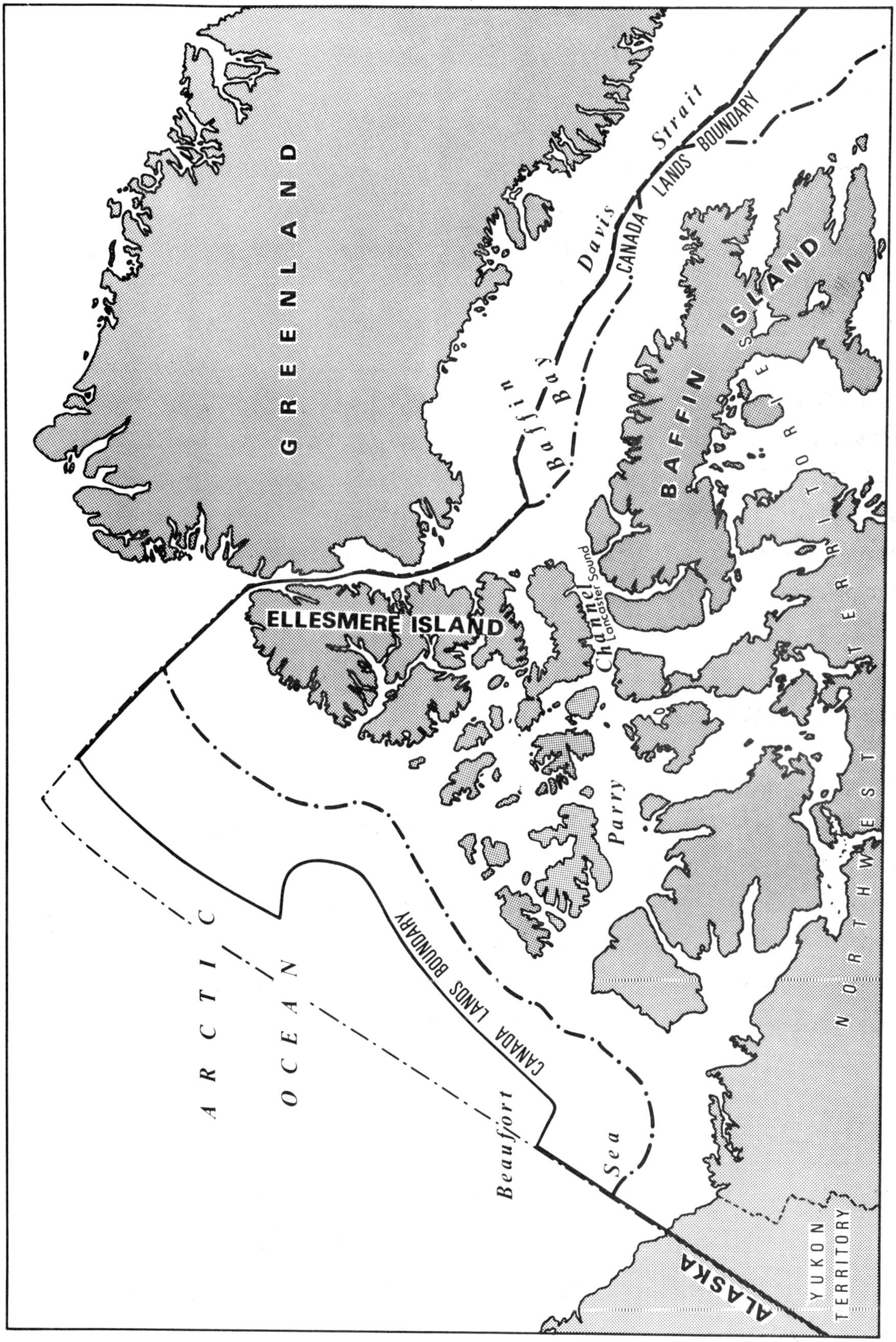

Figure 3. Arctic Zones: The Sector, Canada Lands, and the Arctic Waters Pollution Prevention Act Zone

application may be modified to respect marine areas under the jurisdiction of Greenland.

The Act imposes absolute liability (i.e., complete responsibility) on anyone depositing waste in the special zone, either from ships, other facilities, or from shore, but also sets monetary limits on that liability.[77] A demonstration of financial responsibility up to the limits of liability is required of anyone engaged in any work, undertaking, or transportation involving waste substances.[78]

The government can recover costs and expenses incurred in the mitigation or reduction of pollution, and for the cost of removal of the substance causing damage.[79] Other parties can recover for damage sustained and costs incurred as a result of a deposit. Recovery in both cases is by way of civil action against the party responsible. As there are limits to the imposed liability, claims are ranked, and priority is given to claimants who suffered actual loss or damage.[80]

The Canada Agreement

Due to the proximity of the Canadian Arctic to Alaska, a supplement to the Arctic Waters Pollution Prevention Act, called the Canada Agreement, has been entered into between the Canadian government and its licensees active in the Beaufort Sea (i.e., all interest holders, not merely the relevant operators). The Canada Agreement extends approximately the same remedies available to Canadians under the Act to citizens, corporations, associations, and the governments of Alaska and the United States.[81]

---

77. Id., ss.6, 7. The present limit is 2,000 gold francs (approximately $184) multiplied by the tonnage of the ship, or 210 million gold francs (approximately $19,320,000), whichever is the less - Arctic Waters Pollution Prevention Regulations, C.R.C. 1978, c.354, s.15.
78. Arctic Waters Pollution Prevention Act, R.S.C. 1970 (1st Supp.), c.2, ss.8(1), (2).
79. Id., s.6.
80. Id.
81. Canada Agreement, Art. 3. The Agreement was entered into June 1980, and is unpublished.

Originally, operators engaged in activity under the Beaufort Sea or Davis Strait Drill Ship Programs were liable for an amount fixed by reference to a scale of their activities. Operations in Arctic waters which were not connected with these two programs were also liabile for environmental damage up to limits fixed under separate regulations.[82] These rules have now been changed; the maximum liability for all operators is now $40,000,000.[83] However, under the Canada Agreement "the Operator" is subject to an additional liability to cover damage which might be caused west of the 141st meridian of west longitude, which consitutes the Canada-U.S. boundary for the purposes of the Agreement alone.[84]

The only defence available under both the Act and the Agreement is to show that the claimant's conduct either caused or contributed to the deposit causing the damage. In such cases, the claimant is prevented from recovering to the extent that his conduct contributed to the incident.[85]

The Canada Oil and Gas Act effected changes to the environmental regulations applicable to offshore operators. The government is invested with a broad discretionary power to deal with serious environmental problems in any area to which the Act applies.[86] The interest holders under the Act will be required to contribute to an Environmental Studies Revolving Fund.[87] The Fund is to be maintained at a maximum of $15 million, and contributions are to be fixed by reference to the area held.[88] The minister can order topping-up payments when the Fund falls below $7.5 million.[89]

---

82. Arctic Waters Pollution Prevention Regulations, Amendment, SOR/80-413, s.2; Arctic Waters Pollution Prevention Regulations, Amendment, SOR/79-406, s.2.
83. Arctic Waters Pollution Prevention Regulation, Amendment, SOR/81-447.
84. Canada Agreement, Art. 2.
85. Arctic Waters Pollution Prevention Act, R.S.C. 1970 (1st Supp.), c.2, s.9(1); Canada Agreement, Art. 2.
86. See, for example, remedial orders issued under s.6(b) of the Canada Oil and Gas Act, S.C. 1980-81-82-83, c.81, and the provisions of s.19(4) of the Oil and Gas Production and Conservation Act, R.S.C. 1970, c.O-4 as am. by the Canada Oil and Gas Act.
87. Canada Oil and Gas Act, S.C. 1980-81-82-83, c.81, s.49(4).
88. Id., s.49(5).
89. Id., ss.49(11), (12).

The Act makes provision for the disclosure of confidential information if required in connection with an environmental study, an inquiry into an oil spill, or the status of offshore activities.[90] The Act also prescribes penalties in the event of noncompliance with the environmental provisions; cancellation of a licence is the ultimate sanction for failure to comply with the Canada Oil and Gas Act or the Oil and Gas Production and Conservation Act.[91] Such an order may be appealed to the minister, and may also be subject to judicial review in the federal court.[92]

The Canada Oil and Gas Act also effected amendments to s.19 of the Oil and Gas Production and Conservation Act with respect to liability for oil spills in connection with offshore operations. Operators bear absolute liability for discharges of oil from installations other than ships. There are no limits to liability, and the operator is unable to avail himself of any defence, even that of _force majeure_, sabotage, act of war, or Act of God. Accordingly, an operator on an artificial island may be liable up to the limits to liability, if any, for an oil spill from an artificial island under the Arctic Waters Pollution Prevention Act, the Oil and Gas Production and Conservation Act, and the Canada Agreement (should damage from the spill occur in the United States). Liability will be concurrent under all heads. The full extent of liability is therefore enormous; the overlaps between these various instruments may not have been intentional, and some reforms would therefore be required to rationalize the position.

Regulations have been made under the Oil and Gas Production and Conservation Act concerning safety precautions applicable to drilling, exploration, and production.[93] These provisions also prohibit the abandonment of installations or parts of installations.

In general, offshore operators are required to show evidence of financial responsibility (i.e., ability to meet their maximum liabilities) in the event of environmental damage consequent upon the discharge of oil or debris.[94]

---

90. _Id._, s.50.
91. _Id._, s.56.
92. _Id._
93. Canada Oil and Gas Drilling Regulations, SOR/79-82.
94. Oil and Gas Production and Conservation Act, R.S.C. 1970, c.O-4 as am., s.19(3).

The Ocean Dumping Control Act[95] and regulations made thereunder[96] are designed to prevent or control the dumping of any substance at sea (including dumping on ice cap areas). Dumping a substance at sea, unless authorized by permit issued under the legislation, constitutes an offence except where the dumping is incidental to the normal operation of a ship or aircraft, or is incidental to, or derived from, offshore mineral development operations.[97] This last mentioned provision is rather wide, but would probably not include dumping in connection with transportation of oil and gas, or transportation of substances to or from offshore installations. The Act also prescribes penalties for dumping without a permit.[98]

It is not clear what system will be used to transport offshore hydrocarbons produced in Arctic waters. Meanwhile, the framework for the regulatory regime, which might govern the movement of petroleum from the offshore generally, has been promulgated. The amendments to the National Energy Board Act[99] effected in 1982[100] empower the Governor in Council to direct the National Energy Board (NEB) to assume supervision and control over the movement of hydrocarbons out of "the offshore area", which is defined in terms similar to those employed in the Canada Oil and Gas Act, including Sable Island.[101] No such movements would, in this case, be permitted without a licence issued by the NEB.[102] No such direction has been given to the NEB to date.

The Oil and Gas Production and Conservation Act invests the minister with wide powers of approval over all aspects of the exploration and production of oil and gas.[103] The mode of transportation of any petroleum produced would also be considered as part of the federal Environmental Assessment Review Process for field development.

---

95. Ocean Dumping Control Act, S.C. 1974-75-76, c.55.
96. Ocean Dumping Control Regulations, SOR/75-595.
97. Id., s.2(1)(a), (b).
98. Id., s.13.
99. National Energy Board Act, R.S.C. 1970, c.N-6 as am. For a discussion of the functions and powers of the Board, and its jurisprudence, see Hunt, Constance D. and Lucas, Alastair R., eds., Canada Energy Law Service, Don Mills, Richard de Boo Ltd., 1981, Vol.1.
100. National Energy Board Amendment Act (No. 3), S.C. 1980-81-82-83, c.116.
101. Id., s.87.1(1).
102. Id., s.87.1(2) et seq.
103. Oil and Gas Production and Conservation Act, R.S.C. 1970, c.O-4, s.3.1.

THE COASTING TRADE

The Canada Shipping Act provides that only "British ships" may engage in trade between Canadian ports without a licence.[104] These provisions apply, therefore, to such crafts as supply vessels, tankers, and barges. At the time of writing, application of this part of the Act can be suspended where no Canadian vessels are available to meet a demonstrable need.[105] Canada was a party to the Commonwealth Merchant Shipping Agreement of 1931 whereby the parties granted reciprocal access to each other's merchant ships in respect of coasting trade. Canada was anxious thereby to preserve access for British and Newfoundland ships, but the Agreement outgrew its usefulness when Newfoundland entered Confederation in 1947, and when some "new" commonwealth countries became "flag of convenience" states with extensive merchant fleets. It was felt that Canada was, in effect, permitting some foreign vessels access to the coasting trade on terms more favourable than those available to domestic vessels.

A policy paper released in September 1982[106] outlines proposed changes to the law. Briefly, these would restrict trade between Canadian ports (specifically including installations on the Canadian continental margin as defined in the Canada Oil and Gas Act) to vessels registered in Canada, although foreign vessels may be used temporarily on payment of a fee. This proposal should be viewed in conjunction with plans to extend customs jurisdiction to activities in the offshore.[107] If implemented, the effect of these changes in policy would be to emphasize the jurisdictional connection between petroleum operations on the continental margin and the adjacent coastal state. These activities are seen as industrial enterprises under Canadian jurisdiction as if they were being conducted onshore.

---

104. Canada Shipping Act, R.S.C. 1970, c.S-9, Part XV.
105. Id.
106. Canada. Department of Transport. Marine. New Coasting Trade Policy, a Background Paper, Ottawa, September 1982.
107. Canada. Department of Finance. Extension of Customs Jurisdiction over the Canadian Continental Shelf, a Background Paper, Ottawa, Supply and Services, December 1982.

The definition of a "port" may present certain problems in the context of oil and gas exploration and production, where a variety of offshore loading facilities may be employed. Extending the coasting trade to the continental shelf would prevent non-Canadian vessels from supplying or provisioning offshore installations. This may pose certain problems for rig operators where special vessels or services are required, but Canadian vessels are not available.

The Norwegian government is presently considering legislation which would prohibit non-Norwegian vessels from servicing or supplying provisions for installations on its shelf.[108] This is not necessarily a protectionist measure; the government appears to be concerned that non-Norwegian vessels (which are subject to the jurisdiction of the state of registration) do not comply with Norwegian safety standards. These standards apply not only to protection of the crew, but matters such as storage of waste, and securing debris to be transported ashore from an installation.

COURT JURISDICTION IN OFFSHORE AREAS

The claims made by a nation in respect of the natural resources of the continental shelf beyond the limits of the territorial sea are not territorial, but jurisdictional. Jurisdiction is limited to sovereign rights to explore for and exploit nonliving resources, but the shelf is, in lay terms, outside the nation. Thus, the Beaufort Sea beyond the twelve nautical mile limit claimed by Canada is not part of Canada, notwithstanding that Canada in fact controls resource exploitation there.[109] A consequence of this is that there is no automatic extension of the domestic laws of a nation beyond the outer limit of the territorial sea. Rather, the application of a nation's law beyond territorial limits must be brought about by positive legislative action. Canada has done this, for example, with the Arctic Waters Pollution Prevention Act, the Canada Shipping Act, and in a number of other specific instances, either explicitly or by necessary implication.[110]

---

108. Norwegian Petroleum Directorate, personal communication.
109. See, however, the recent decisions in R. v. Tootalik E4-321 (1970), 71 W.W.R. 435 (N.W.T.T.C.), and B.P. Exploration Co. (Libya) Limited v. Hunt, ([1981] 1 W.W.R. 209 (N.W.T.S.C.)); the impact of these decisions on the seaward boundary of the Northwest Territories is discussed infra.
110. See de Mestral, supra, note 31.

Canadian law may, as a matter of both Canadian and international law, have extra-territorial effects. Thus, Canadian courts have a residual jurisdiction over, for example, a crewman on a Canadian ship who commits an offence on that ship wherever it may be.[111]

To what extent has Canada extended the jurisdiction of its courts to deal with matters that arise on the continental shelf? It is clear that the Canada Oil and Gas Drilling Regulations[112] apply on drilling units on the shelf. It is also clear that Canadian law will apply on Canadian ships or mobile installations which may be classified as Canadian ships on the Canadian shelf.[113] The Income Tax Act, for example, applies for certain purposes in the offshore.[114]

With these exceptions, Canadian law does not apply to the offshore, and Canadian courts would lack jurisdiction to try issues of, for example, tort, contract, or criminal law arising from acts or omissions on offshore installations which would not otherwise bear the character of "Canadian ships".[115]

There appears to be a gap, therefore, in the domestic integrity of Canada's offshore claims. This gap does not undermine the validity of these claims, but full control of offshore operations, in the fullest sense of the term, is difficult to enforce. This is a gap which could be easily plugged by the appropriate legislation

---

111. R. v. Anderson (1868), XI Cox's Criminal Law Cases 198 (C.A.). Although this is an English case, it applies a principle which has been received into Canadian law. See also the Interpretation Act, R.S.C. 1970, c.I-28, s.8.
112. Canada Oil and Gas Drilling Regulations, SOR/79-82.
113. Convention on the High Seas, 29 April 1958, in U.N. Doc. A/CONF 13/L.53. Although not ratified by Canada, much of this Convention is a codification of existing international law.
114. Income Tax Act, R.S.C. 1952, c.184 as am. by S.C. 1970-71-72, c.63, and subsequently amended, s.255.
115. See Law of the Sea Convention, Art. 60(2); Lawrence L. Herman, "The Need for a Canadian Submerged Lands Act: Some Further Thoughts on Canada's Offshore Mineral Rights Problems", LVIII Can. Bar Rev. 518 (1980).

enacted by Parliament. As experience in the North Sea has shown, the consequences of this omission can be very serious.[116]

Some of these jurisdictional issues can arise in a unique context in the Arctic. The jurisdictional boundaries of the Northwest Territories have been in question following two recent decisions: R. v. Tootalik E4-321,[117] where courts have taken jurisdiction over an act which took place on offshore ice, and B.P. Exploration Co. (Libya) Limited v. Hunt,[118] where the jurisdiction of the court was held to include the seabed forty to fifty miles offshore.

It may be argued that in these two cases the court was willing to do justice at the expense of certainty over boundaries. In Tootalik, a hunter had killed a bear on offshore ice. Had the incident occurred "onshore", it would have been illegal under Northwest Territories legislation. The judge merely extended the boundary to include the incident. In B.P. v. Hunt, creditors were seeking to secure assets which were offshore mineral holdings. The court moved the boundary seaward to bring the assets within its jurisdiction for the purposes of the action. This represents a departure from usual Canadian practice, which regards the low-water mark as the limit of court jurisdiction.

Clearly, some clear legislative guidance is required to prevent the *ad hoc* development of court jurisdiction. The multifarious nature of offshore monitoring and supervision must ideally take place within a settled legal regime.

---

116. After the loss of the rig Sea Gem in 1965, which was operating on the British continental shelf, it was discovered that U.K. law had made no provision for the registration of the deaths of workers, nor could a fatal accident inquiry be convened to inquire into the circumstances of the accident; legislation was subsequently introduced in an attempt to plug these gaps in the law.
117. R. v. Tootalik E4-321 (1970), 71 W.W.R. 435 (N.W.T.T.C.).
118. B.P. Exploration Co. (Libya) Limited v. Hunt, [1981] 1 W.W.R. 209 (N.W.T.S.C.). For a comment on this case see W. Wylie Spicer, "Some Admiralty Issues in Offshore Oil and Gas Development", XX Alta. L. Rev. 63 (1982).

CONSTRUCTION AND MAINTENANCE OF OPERATIONAL FACILITIES IN OFFSHORE AREAS

Canadian law does apply offshore for certain limited purposes. Regulations apply on offshore facilities covering such matters as the proper conduct of drilling operations and safety. As the categories of installations in the Canadian sea proliferate (mobile drilling units, fixed drilling units, drill ships, artificial islands), it will also be necessary for the domestic legal system to define the juridical status of each.[119]

In Canada, jurisdiction over the construction and siting of artificial islands or installations in Canadian Arctic waters has been assumed by the Department of Indian Affairs and Northern Development. The Department appears to base jurisdiction over the former on the Territorial Land Use Regulations promulgated pursuant to the Territorial Lands Act. Section 2 of the Act defines territorial lands to include land in the Northwest Territories or Yukon Territory that are vested in the Crown or of which the government of Canada has power to dispose.

A further elaboration of this definition is found in the Oil and Gas Land Order No. 1 - 1961[120] which seeks, through DIAND, to regulate the grant of permits for development of areas which, in the minister's opinion, fall into the category of "seacoast waters", a term which is not defined in the Order.

There is, therefore, no discrete, detailed legal regime for the construction of artificial islands. DIAND has wide discretionary powers when considering applications from prospective builders. It should be noted, however, that the Territorial Lands Act, like every other Canadian statute, is presumed _not_ to apply outside "Canada" (that is, seaward of the low-water mark or the baselines), unless specifically stated. This Act makes no such provision; it is therefore subject to the presumption against extra-territorial application. Hence, DIAND's assumption of jurisdiction rests on a somewhat insecure legal foundation.

---

119. A consequence of failure to do this was seen from the unsatisfactory decision in _Dome Petroleum Ltd._ v. _Hunt International Petroleum Co._, [1978] 1 F.C. 11.
120. Oil and Gas Land Order No 1 - 1961, SOR/61-461.

Another related jurisdictional problem arose in late 1982, when the government of the Northwest Territories announced its intention to tax artificial islands in the Beaufort Sea, a decision which was welcomed neither by the federal government nor by offshore operators. Clearly, a Canadian authority would be empowered to tax such structures if intra vires its powers. There are really two questions here: first, does the government of the Northwest Territories enjoy such powers, and second, are the islands in question within the jurisdiction of these powers?

This disputed claim to jurisdiction (and others which may follow it) arises from lacunae or ambiguities in Canadian constitutional law. The federal-provincial jurisdictional dispute arose from the absence of a specific constitutional provision investing rights to explore for and exploit offshore oil and gas in one level of government or the other.[121] The Canadian Constitution antedates the discovery of these resources and the formulation of the international legal regime concerning them.

But just as international legal norms were relevant to a determination of the east coast dispute, so are they relevant to determining the powers of the government of the Northwest Territories to levy taxes on artificial islands lying off its coast. From the domestic viewpoint, the question is ultimately one of fixing the boundaries of the Northwest Territories and hence the jurisdiction of its government in the offshore, bearing in mind the changing juridical character of offshore zones, which will be discussed in the section "The Status of the Waters of the Canadian Arctic Archipelago". While it is hardly surprising that a government should attempt to take jurisdiction to resolve an ambiguity, taxation power can only be exercised on the basis of firm jurisdiction. It is not enough to show that there is a lacuna - a loophole - in the law in such circumstances.

It must follow from the discussion above that the Northwest Territories government can only levy such taxes on islands in the Beaufort Sea if it clearly enjoys extra-territorial jurisdiction, since such islands are "outside" Canada. The

---

121. Similar disputes have also arisen in the United States and Australia, where they have been settled for all practical purposes. For a comment on the U.S. solution, and the present state of the Canadian dispute, see Townsend Gault, Petroleum Operations on the Canadian Continental Margin, supra, note 27.

jurisdictional position vis-a-vis facilities in waters between the Arctic islands is different; these issues will be discussed in the section on "Offshore Facilities".

Jurisdiction over the construction and siting of installations such as oil rigs, production platforms and the like, is more straightforward. These matters are regulated by the Canada Oil and Gas Drilling Regulations.[122]

122. Canada Oil and Gas Drilling Regulations, SOR/79-82.

THE INTERNATIONAL LEGAL CONTEXT OF CANADA'S CLAIMS IN THE ARCTIC

The purposes of this part are twofold. First, the specific domestic Canadian claims identified in the previous part are set in the context of international law as it is emerging, and their validity is tested. Second, Canada's broader international claims are discussed and the relevant principles of international law identified. The second purpose is somewhat difficult to meet satisfactorily, however, as it is not always possible to identify the nature and extent of Canadian claims. This is, perhaps, one of the consequences of the inherently evolutionary nature of the international law-making process, identified in the first part.

APPLICATION OF INTERNATIONAL LAW TO ARCTIC WATERS

The work of the United Nations conferences on the law of the sea has already been referred to. While the five Conventions[123] which have emerged from this process have been accepted, in large measure, as a codification and development of the international law of the sea, it is necessary to consider whether that law is applicable in Arctic waters.

Some jurists have sought to make a case for the Arctic as an area which is legally *sui generis* or unique. They argue that to regard Arctic waters as waters like any other ignores the special characteristics of the region. It is not difficult to grasp the reasons behind this argument. As will become apparent in the discussion below, the nature and extent of jurisdiction in the Arctic is complicated by a number of factors: the presence of a permanent ice area as opposed to land, uncertainty as to boundaries - the legacy of the sector theory, and the delicate nature of the Arctic environment. There is also a feeling that one of the last wilderness areas on Earth should remain inviolate, at least so far as is possible. These unique considerations, it is argued, lend support to unique treatment from the legal viewpoint. Established concepts of law, such as freedom of the seas,

---

123. The Convention on the Territorial Sea and Contiguous Zone, the Convention on the High Seas, the Convention on the Conservation of the Living Resources of the High Seas, the Convention on the Continental Shelf, and the United Nations Convention on the Law of the Sea.

should not apply. If this view is to be accepted, it may also be argued that the domestic legal regime should similarly distinguish between Arctic waters and other marine areas within the jurisdiction of a state.

The alternative viewpoint is that the law of the sea should apply in the Arctic, but that special rules should be promulgated to take account of, for example, the Arctic environment. The law of the sea and the domestic legal regime would otherwise apply.

To ascertain the legal position, and to distinguish that position from the rival claims of the protagonists in this argument, it is suggested that the practice of states must be closely examined. Canadian practice and claims have already been outlined above, but the paucity of practice in Arctic areas generally precludes the identification of a clear rule or course of conduct. Such evidence as there is indicates that the latter view is adhered to. Canada, the United States, and the Soviet Union are conducting oil and gas activities in Arctic waters. Plans for the transportation of hydrocarbons by sea are being mooted. In the case of Canada, a special regime for the prevention of pollution has been promulgated, and international law is moving to recognize the validity of such measures - see the section "Environmental Protection". Accordingly, it is necessary to examine the present state of the law of the sea, including special rules for Arctic waters, and to compare these emerging norms to Canada's claims with respect to oil and gas activities in northern waters.

It will be recalled that the beginnings of a new principle of international law may frequently be traced to a single claim, laid by one state unilaterally and later followed by other states to the point where it becomes widely accepted. This is not always a straightforward process, and states may later retreat from a claim if support is not forthcoming in the form of similar claims by others. Even where well established principles are applied to particular circumstances, there may be a similar phenomenon of advance and retreat, as the acceptability of a claim is tested. Thus, the position of a state with respect to a particular claim may not be easily identifiable all the time. In such cases, resort may be had to what that nation has said and done in the past. These considerations have particular force when applied to some of Canada's international claims in the Arctic.

The following discussion will refer to some extent to relevant provisions of the Law of the Sea Convention. It must be emphasized, however, that those provisions may or may not be subsequently established as rules of international law. Even when a convention is finally drafted and open for signature, it will enter into force only when ratified (ratification is a process beyond mere signature) by the required number of states, and will bind them alone until such time as the treaty may be said to have received near universal acceptance.[124] Until this happens, Canada can only be bound by ratification of the Convention.[125]

CLAIMS TO THE ARCTIC ISLANDS

International law recognizes claims to territory on several grounds. An unoccupied island, not already claimed by another state, may come under the jurisdiction of the first occupant to make a claim. Normally, a claim must be backed up by physical occupation, though where this is impossible due to the nature of the island, all that is required is some official indication of the claim, such as maintenance of navigational aids, or inclusion on official maps and charts.[126]

---

124. It is possible for international customary law to evolve to a point where a convention becomes merely declaratory of existing law. This has happened, to a large extent, in the case of the Convention on the High Seas.
125. In Canadian practice, ratification is accorded by the Governor General in Council in exercise of the royal prerogative. Ratification does not mean that the treaty is automatically incorporated into Canadian law.
126. For a discussion of the legal principles applicable here, see the Island of Palmas Case, II U.N.R.I.A.A. 829 (1928); the Clipperton Island Case, I U.N.R.I.A.A. 1105 (1931); the Eastern Greenland Case, 1933 P.C.I.J. Reports, Series 1 - Ref. S A/B No. - 77. See also Bowett, Derek W., The Legal Regime of Islands in International Law, New York, Oceana Publications Inc., 1978, Chapter 2.

THE TERRITORIAL SEA

The breadth of territorial sea recognized at international law has been a subject of controversy and uncertainty for some years.[127] The Convention on the Territorial Sea and Contiguous Zone set a maximum limit of twelve nautical miles by implication only;[128] no specific limit was set in the Convention. Certainly claims in excess of twelve nautical miles would then be, and probably still are, contrary to international law. The Canadian twelve nautical mile limit is therefore justified.

Canada has yet to promulgate an order drawing straight baselines around the islands of the Arctic archipelago. Accordingly, in both international and Canadian law, the baseline is constituted by the low-water mark. Thus, lines enclosing the Arctic islands as a group have not been drawn.

Reference has already been made to Canada's jurisdictional claims made under fisheries legislation and the Arctic Waters Pollution Prevention Act. Assessment of the validity of these claims may be carried out on their respective merits, and this is attempted in succeeding sections. But Canada's claim to jurisdiction over the waters of the Arctic archipelago overlaps with and complicates these claims, for this is a claim to full sovereignty, unhampered by the right of innocent passage.

Before considering a claim based on historic title, I propose to examine the present state of an older Canadian claim, as constituted by the sector theory, and, subsequently, the emerging rules of international law concerning archipelagos and straight baselines.

Directions issued by the Minister of the Environment following recommendations made under the EARP process could clearly apply to vessels flying the Canadian

127. A three nautical mile limit was generally accepted in 1950, by 1958 this was somewhat less certain, though probably the rule, but claims of up to twelve nautical miles may now be regarded as the norm; see Law of the Sea Convention, Art. 3.
128. By setting the outer edge of the Contiguous Zone (a zone seaward of, but adjacent to, the Territorial Sea) at twelve nautical miles, the Convention implied this extent for the Territorial Sea in the absence of a set limit.

flag. There would appear to be jurisdictional problems in the way of enforcement *vis-a-vis* foreign vessels on the high seas. Foreign vessels within Canadian territorial waters would, *prima facie*, be exempt, unless it could be held that noncompliance would be construed as a violation of the right of innocent passage. This, it is submitted, could only be maintained where a real threat to the environment was apprehended through noncompliance.[129]

CONTROL OF RESOURCE EXPLORATION AND PRODUCTION

LIVING RESOURCES

Some coastal states have been urging since the first world war that state jurisdiction over the living resources of the high seas out to the edge of the continental shelf is justified by the need to conserve fish stocks.[130] Similar motives prompted Central and South American countries to claim 200 nautical mile territorial seas in the years following the Truman Proclamation. The Convention on the Living Resources of the High Seas failed to strike an acceptable balance between the wishes of the conservative maritime states and those countries which supported increased coastal state jurisdiction over fishing. This failure did not deter states such as Iceland, and a series of diplomatic victories through which that country's exclusive fishing zone was extended to 200 nautical miles virtually forced other coastal states to follow suit. These claims on the Exclusive Economic Zone are codified in Part V of the Law of the Sea Convention.

---

129. Article 14(4) of the Territorial Sea Convention provides that "(p)assage is innocent so long as it is not prejudiced to the peace, good order or the security of the Coastal State". In 1969, the Institute of International Law debated measures to prevent pollution of the seas. It finally proposed that "States have the right to prohibit any ship that does not conform to the standards set out in accordance with the preceding articles for the design and equipment of the ship ... from crossing the territorial seas ... and Contiguous Zone and from reaching their ports" - in: Measures Concerning the Prevention of Pollution of the Sea, Article VI, Geneva, Institute of International Law, 1969. It should be remembered that this is only the opinion of the Institute, important as that may be.
130. Senor De Buren, later to become Director General of Spanish Fisheries, made such a proposal in 1916; see League of Nations Document C.196. N70, 1927 at 63. The Argentinian authorities made a similar proposal at about the same time.

Coastal states may exercise within the zone exclusive rights for the purposes of exploring for and exploiting all natural resources, living and nonliving. The zone extends 200 nautical miles from the baselines. The state may not claim ownership either over the waters of the zone, nor over the resources thereof. It enjoys within the zone only an exclusive right to take the resources.

Canada claimed a two hundred nautical mile fishing zone in 1977.[131] The zone is not measured from the baselines, but by using the basepoint method described above. This claim is similar in terms to those which have been made by many coastal states since the failure of the Convention on the Living Resources of the High Seas to settle the issue between the conservative and expansionist states, and which in turn prepared the way for the establishment of the concept of the Exclusive Economic Zone as a rule in international law.[132]

NONLIVING RESOURCES

It is settled law that jurisdiction over exploration for and production of the natural resources of the seabed and subsoil of the continental shelf vests in the coastal state.[133] The Convention on the Continental Shelf, to which Canada is a party, established a regime which accorded only those rights strictly necessary to enable a country to supervise these activities. These rights are to be exercised for the purpose of exploration for and exploitation of the natural resources (including minerals) of the seabed and subsoil.[134] The Convention set the seaward limits to state jurisdiction at the 200 metre isobath, or the limits of exploitability, whichever is the greater.[135]

---

131. Effective January 1, 1977. The zone in the Arctic is established by the Fishing Zones of Canada (Zone 6) Order, C.R.C. 1978, c.1549.
132. Canada's claim is more or less in line with similar moves on the part of Mexico, India, the U.S., Norway, the United Kingdom, France, Spain, Portugal, the U.S.S.R., Poland, East Germany, and Japan; these claims were made at the same time as, or within a few years of, Canada's own.
133. <u>North Sea Continental Shelf Cases,</u> [1969] I.C.J. Rep. 3 at 22.
134. Convention on the Continental Shelf, Art. 2(1).
135. This formula was used to determine the extent of application of the Oil and Gas Production and Conservation Act, R.S.C. 1970, c.O-4, in offshore areas, and would have formed a part of the definition of "Canada Lands" in Bill C-20, <u>supra,</u> note 26.

This formula does not fix a limit to the seaward extent of state jurisdiction; determination of fixed limits to national jurisdiction in the offshore was one of the purposes for which UNCLOS III was convened. Many states and jurists wished to see the establishment of such rules and the consequent recognition of the concept that marine areas beyond the limits of state jurisdiction and their resources, were part of the common heritage of mankind, not subject to the unilateral jurisdiction of any one state. This principle proved unacceptable to the United States government and some western European countries when the Convention was opened for signature and ratification in December 1982.

The new regime for nonliving resources is considerably more detailed than the old.[136] The coastal state enjoys only sovereign rights over the shelf for the purposes of exploration for and exploitation of the resources of the seabed and subsoil.[137] The rules in the Law of the Sea Convention for determining the edge of the continental margin beyond the 200 nautical mile limit permit states to apply one of several alternative methods, although the final result is to be approved by a proposed Commission on the Limits of the Continental Shelf.[138] At all events the absolute limit is either 350 nautical miles from the baselines, or 100 nautical miles from the 2,500 metre isobath.[139] The status of the superjacent waters as high seas is preserved, and the principle of the freedom of the seas maintained; development of natural resources must take place in accordance with this principle.[140]

---

136. The regime for nonliving resources is covered by both Parts V (Exclusive Economic Zone) and VI (Continental Shelf) of the Law of the Sea Convention. The rules for determination of the seaward limits beyond 200 nautical miles are laid out in Article 76 which forms part of Part VI.
137. Law of the Sea Convention, Art. 77.
138. Law of the Sea Convention, Art. 76(8). For details of the Commission, see Annex II of the Convention.
139. Id., Art. 76(5).
140. Id., Art. 78.

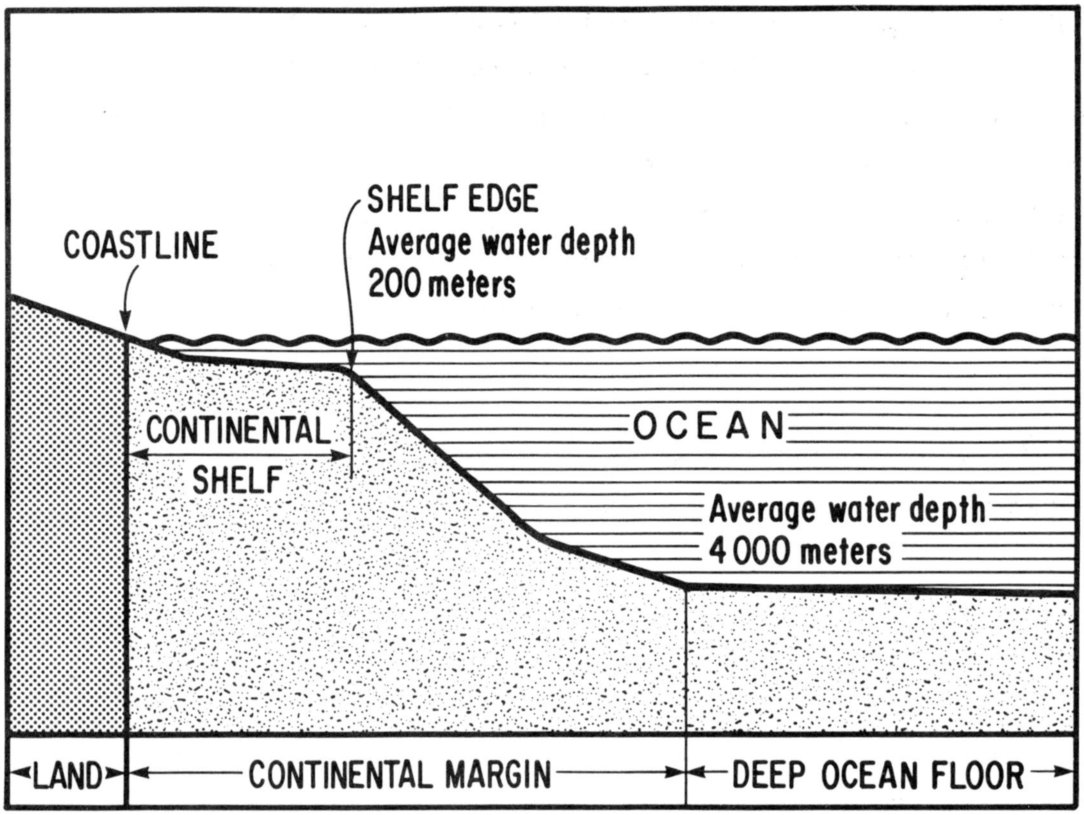

Figure 4. The Continental Margin

ENVIRONMENTAL PROTECTION

The last twenty years have seen the introduction and (to a greater or lesser extent) acceptance of measures designed to prevent oil pollution of the seas. These measures are implemented by way of treaties, of which the most important in this context are those concerning: prohibition of dumping of oil contaminated water from tankers,[141] action which may be taken with respect to marine

141. International Convention for the Prevention of Pollution of the Sea by Oil, 1954, in New Directions, supra, note 58, Vol. II at 580; October 15, 1971 amendments in New Directions, Vol. II at 489.

casualties likely to pose a pollution threat,[142] tanker design,[143] liability for pollution damage as a result of oil emanating from ships,[144] and the establishment of a transnational regime for ensuring the payment of claims for compensation in respect of pollution and clean-up costs damage.[145] These treaties have been enacted into municipal law by many petroleum producing countries.[146]

Some jurisdictions have enacted legislation concerning the conduct of offshore operations, as in prohibitions against littering of the sea bottom by debris.[147] Tanker owners and the oil industry have, in response to government pressure, produced schemes concerning liability and compensation for oil pollution damage from ships.[148] The North Sea states have encouraged their operators to develop a

---

142. Convention Relating to Intervention on the High Seas in Cases of Oil Pollution Casualties, 26 November 1969, in IX Int'l Leg. Mat. 25 (1970); reprinted as am. in 1962, 1969, and 1971 in New Directions, supra, note 58, Vol. II at 592.
143. International Convention for the Prevention of Pollution from Ships, November 2, 1973, in XII Int'l Leg. Mat. 1319 (1973); New Directions, supra, note 58, Vol. IV at 345.
144. International Convention on Civil Liability for Oil Pollution Damage, November 29, 1969, in IX Int'l Leg. Mat. 45 (1970); New Directions, supra, note 58, Vol. II at 602.
145. International Convention on the Establishment of an International Fund for Compensation for Oil Pollution Damage, December 18, 1971, in XI Int'l Leg. Mat. 282 (1972); New Directions, supra, note 58, Vol. II at 611.
146. These treaties set out standards which, to be effective, must be implemented in domestic law by states party thereto.
147. The problem of debris discharged from installations and supply vessels has reached such proportions in Norway that special regulations have been introduced to deal with it, the Provisional Regulations Concerning Littering and Pollution Caused by Petroleum Activities, 1979. These require debris to be destroyed safely or transported ashore in approved containers, such movements being logged and otherwise supervised.
148. Tanker Owners Voluntary Agreement Concerning Liability for Oil Pollution, January 7, 1969, in VIII Int'l Leg. Mat. 497 (1969); New Directions, supra, note 58, Vol. II at 641, and the
Oil Companies Contract Regarding an Interim Supplement to Tanker Liability for Oil Pollution, January 14, 1971, in X Int'l Leg. Mat. 137 (1971); New Directions, supra, note 58, Vol. II at 646.

similar scheme for installations.[149] The U.S.A.[150] and Norway[151] have enacted schemes to compensate fishermen suffering as a result of offshore activity; the U.K. on the other hand, has encouraged the oil industry itself to do so.[152]

The Law of the Sea Convention contains provisions which seek to ensure the proper conduct of offshore operations on the continental shelf and indicate the supervisory role marked out for the coastal state. Article 234 of the Law of the Sea Convention permits the coastal state to adopt and enforce nondiscriminatory laws to prevent, reduce, or control marine pollution from vessels in ice-covered areas within the Exclusive Economic Zone. Climatic conditions must be severe, and the ice must constitute a hazard to navigation; the threat to the ecological balance of the area posed by pollution must, therefore, be acute.

This provision is the first explicit support for Canadian policies with respect to pollution control in the Arctic. The permitted area of jurisdiction, the Exclusive Economic Zone, extends two hundred nautical miles from the baselines, and Canada has already laid claim to a 200 nautical mile fishing zone in addition to the claim to a continental shelf of at least that width.[153] The zone under the Arctic Waters Pollution Prevention Act is one hundred nautical miles "from the

---

149. Offshore Pollution Liability Agreement, September 1974, in XIV Int'l Leg. Mat. 1409 (1974), annotated in Daintith, Terence and Willoughby, G.D.M., eds., A Manual of United Kingdom Oil and Gas Law, London, Oyez Publishing Ltd., 1977 at 488.
150. The Federal Fishermen's Contingency Fund (established under the Outer Continental Shelf Lands Act Amendment of 1978, 43 U.S.C., § 1841 et seq.; Rules and Regulations of the Fund appear in the Federal Register, July 2, 1980, § 296.1). The Fund provides compensation in certain circumstances for U.S. commercial fishermen for damage sustained in waters superjacent to the U.S. shelf beyond the territorial sea, which is under the jurisdiction of the coastal states.
151. Regulations for a Temporary Compensation Arrangement for Fishermen in Respect of Loss Caused by Refuse from Petroleum Activity, Royal Decree of February 7, 1981.
152. This scheme, which has no legislative effect, is implemented by the United Kingdom Offshore Operators Association (UKOOA) but claims are adjudicated solely by representatives of the fishing industry.
153. Law of the Sea Convention, Part V. The coastal state may exercise exclusive control over the management and conservation of fishing in this zone (200 nautical miles from the coast), and has, therefore, an interest in oil pollution prevention in such an environmentally sensitive area.

nearest Canadian Land";[154] it seems on the face of it that when the Law of the Sea Convention enters into force, Canada will be able, at the very least, to assert the jurisdiction claimed in the Act with respect to ships flying the flag of other states party to the Convention, provided, of course, that no valid reservation to Article 234 has been made by them.

TRANSPORTATION OF OIL AND GAS

TRANSPORTATION BY PIPELINE

Submarine pipelines have been assimilated into the legal principles applicable to cables - the coastal state may construct and maintain pipelines anywhere (within reason) on the seabed for the delivery of oil and gas onshore. Furthermore, the Convention on the Continental Shelf forbids a state to prohibit other states from laying cables or pipelines on its continental shelf without good reason.[155] In the North Sea, however, Norway agreed to bury that part of the Ekofisk Pipeline (en route from the Ekofisk field to Emden, West Germany) which would overlay the Danish sector of shelf following a request from the latter. It is not clear whether Norway was, or felt itself to be, under any legal obligation to comply.

The Convention also contains a general prohibition against activities on the shelf which interfere unreasonably with fishing and navigation.[156] Burying a pipeline will reduce the risk to a fisherman's nets, but few North Sea pipelines and few submarine cables are buried. Some countries forbid fishermen to work in sea areas adjacent to pipelines.[157]

These principles are amplified by the Law of the Sea Convention. States would be permitted to impose conditions on pipelines entering its territory, territorial seas, or used in connection with the exploitation of any part of the natural resources of its continental shelf.[158]

---

154. Arctic Waters Pollution Prevention Act, R.S.C. 1970 (1st Supp.), c.2, s.3.
155. Convention on the Continental Shelf, Art. IV.
156. Id., Art. V.
157. The United Kingdom and Norway, for example.
158. Law of the Sea Convention, Art. 79.

TRANSPORTATION BY TANKER

International law and maritime law have developed a detailed legal regime for the transportation of oil by tanker. This regime deals with such matters as appropriate shipping routes for tankers, dumping of oil and contaminated water at sea, questions of liability and compensation resulting from major oil spills, and standards of construction for oil tankers.[159]

Canada has attempted to develop the law with respect to sensitive environmental areas through the Arctic Waters Pollution Prevention Act, discussed in detail above. Canada was clearly aware of the difficulties attending the establishment of a new principle, hence the intimation to the United Nations that Canada would not accept the compulsory jurisdiction of the International Court of Justice on any matter relating to the Act. The Prime Minister's letter indicated that Canada was not willing to lay herself open to international litigation which might jeopardize the protection of her sensitive Arctic environment.

It is possible to test the validity of domestic legislation against already existing international legal standards. But it is not always _necessary_ in all cases for the international regime to have laid the foundations for developments in the domestic system. Accordingly, the Arctic Waters Pollution Prevention Act is not invalid _vis-a-vis_ foreigners and foreign vessels on the high seas, but within the special zone, solely because there is arguably little direct support for it in international law. If the waters within the zone outside territorial limits were commonly used by vessels of many nations, silence on the part of those nations might, in time, imply acquiescence and therefore consent. The few foreign vessels passing through the zone have not contravened the terms of the Act.

The validity of such an exercise of jurisdiction as a matter of international law can be shown in a number of ways. Express approval on the part of other states, especially those likely to be affected, would certainly suffice; the U.S.S.R. alone

159. There are a number of recent works which deal comprehensively with the legal regime: M'Gonigle, R. and Zacher, M.W., _Pollution, Politics and International Law: tankers at sea_, Berkeley, University of California Press, 1979, examines the law from a Canadian viewpoint; a more general treatment can be found in Abecassis, D.W., _The Law and Practice Relating to Oil Pollution From Ships_, London, Butterworths, 1978.

has explicitly approved of the legislation. Absence of protest by states likely to be affected might, over a period of time, similarly indicate acceptance as opposed to approval.

The waters within the zone fall into a number of juridicial categories - internal waters, territorial waters, the 200 nautical mile fishing limit, and high seas. However, those waters do not bear the character of an international waterway, though traffic is expected to increase within the next few years as a result of oil and gas activities. Realization of the proposed extension of the Customs Act and implementation of the new coasting trade policy (discussed in the text at note 74) would ensure that these vessels would fly the Canadian flag.

The United States, which is, with Canada, the state most directly affected by the Act, has specifically objected to the extension of jurisdiction claimed therein and has made it clear that the United States does not recognize and will not accept Canadian jurisdiction insofar as it seeks to apply that jurisdiction to United States vessels or nationals.[160] The official United States position is therefore clear; de facto compliance with the Act by U.S. ships may affect that position as a matter of law. The attitude of other states is correspondingly less clear.

THE COASTING TRADE

The coasting trade would not appear to raise any problems in international law. It is clearly open to the coastal state to enforce municipal law on offshore facilities and within the safety zones around them.[161] The provisions of the Canada Shipping Act dealing with the coasting trade could be extended to offshore installations in this way.

Canada would, therefore, be able to enforce safety standards concerning the storage of waste products or debris on vessels and the taking of such refuse ashore

160. U.S. press release, 15 April 1970, reproduced in: Canada. House of Commons. Debates, Appendix "A" at 5923. See also Louis Henkin, "Arctic Anti-Pollution Legislation: Does Canada Make or Break International Law?", 65 Am. J. Int'l L. 131 (1971).
161. Convention on the Continental Shelf, Arts. 5(2), (4).

for proper disposal. Jurisdiction over foreign vessels on the high seas carrying waste from a Canadian installation (whether to the mainland of Canada or not) would be difficult to enforce. Canada could enforce legislation preventing dumping of waste at sea by extending coverage of coasting trade legislation to offshore facilities, a measure which is presently being considered by other countries.[162]

THE STATUS OF THE WATERS OF THE CANADIAN ARCTIC ARCHIPELAGO

There are several issues here, not necessarily interconnected: the status of the sector theory, the Canadian Arctic islands as an archipelagic extension of the mainland, historic title, the law relating to international straits, and finally how the issues arising apply to a determination of the status of the Northwest Passage. Each will be examined in turn.

THE SECTOR THEORY

As indicated in the discussion on the Arctic Waters Pollution Prevention Act, Canada has from time to time perceived its jurisdictional needs in the Arctic islands as being unique. Accordingly, it has taken the initiative in attempting to establish new principles of international law which would accommodate these perceived needs, the Arctic Waters Pollution Prevention Act itself being the classic illustration. But in addition to this specific initiative, Canada has sometimes sought to establish a more broadly based claim that would, if accepted, accord to it a rather general jurisdiction over the whole area. In so doing, Canada has, from time to time, embraced "the sector theory" and indeed, has occasionally been a leading proponent thereof.

This theory claims that the sovereign jurisdiction of the state extends to all land and sea north of the Arctic Circle bounded by meridians drawn from the easterly and westerly extremities of the state from the Circle to the North Pole.

162. Norway is presently considering such legislation; personal communication from Norwegian Petroleum Directorate personnel.

Application of the theory to the North Pole results in a division of the polar area north of the Circle into a series of irregular segments. The theory has been alternatively embraced and neglected by Canadian governments, and now seems dormant, insofar as it was intended to claim Canadian sovereignty to the North Pole.[163] In his seminal work, The Law of the Sea of the Arctic, Professor Pharand charts the course of Canadian government opinion concerning Arctic sovereignty.[164]

The present government position was set out by Prime Minister Trudeau in 1969: "I believe the sector theory applies to the seabed and the (continental) shelf. It does not apply to the waters. The continental shelf is of course under Canadian sovereignty - this is the seabed, but not the waters over the shelf."[165] This statement may be interpreted as signifying a claim to the seabed of the sector beyond the limits of the Canada Lands, and a claim to ownership over the seabed of the sector, rather than sovereign rights for the purpose of mineral resource exploration and development. It is also possible that the Prime Minister was employing terms such as "sovereignty" rather loosely; that term is often misused to describe a country's claims to the continental shelf. At all events, nothing in Canadian law makes a claim to ownership over the seabed of the sector; as has already been remarked, there is no vesting provision in the Canada Oil and Gas Act.

From the Canadian point of view, it would seem that whatever modern utility the theory possessed was almost exhausted when it became clear that title to the Arctic mainland of Canada and the archipelago was beyond dispute; Canada's claim to a two hundred nautical mile fishing zone, mineral rights on the continental margin, and pollution control in the Arctic have substantially replaced the claim made by the sector theory. Contemporary government policy seems to rely on traditional concepts such as effective occupation as a basis for sovereignty in the North. Certainly, some contemporary government maps of "Canada" still indicate the lines of the sector, the 114th line of longitude to the west, and the northern continuation of the Canada-Greenland shelf boundary to the east. The

---

163. The practice of the U.S.S.R. with respect to the Northwest Passage indicates that that country still adheres to the theory.
164. Pharand, supra, note 75 at 134 to 144.
165. Canada. House of Commons. Debates, Vol. VI at 6396 (1969).

maritime boundary with the United States in the Beaufort Sea remains to be delimited. The utility of the theory is not completely exhausted.

In this connection one must note Professor Auburn's comment[166] on the decision in the Tootalik case, to the effect that the court seems to have taken jurisdiction by virtue of an application of the sector theory. This interpretation is important when considering the position of offshore facilities, including artificial islands, in Arctic waters, and is discussed in the section "Offshore Facilities".

THE CANADIAN ARCTIC ISLANDS AS AN ARCHIPELAGIC EXTENSION OF THE MAINLAND

Archipelagic states, such as Indonesia, the Philippines, and Fiji, have long felt that the rules for baselines around islands as reflected in the 1958 Convention on the Territorial Sea took no account of special needs of states wholly composed of one or more archipelagos. Simple application of these rules would produce enclaves of jurisdiction around individual islands, but preserve the status of the waters between islands as high seas. The archipelagic states felt that this left them unable to exercise jurisdiction for internal security purposes and to protect fishing stocks in these waters. The Law of the Sea Convention reflects this concern by permitting such states to draw straight baselines enclosing the islands of the archipelago.[167]

It is from these baselines that the breadth of offshore jurisdictional zones are to be measured, but the right of transit passage in recognized sea lanes through the archipelagic waters, which would otherwise be inland waters and where no right of innocent passage exists, is specifically upheld.[168]

Introducing the Bill to amend the Territorial Sea and Fishing Zones Act,[169] the Secretary of State for External Affairs stated that whether or not Canada chose

---

166. F.M. Auburn, "International Law - Sea-Ice - Jurisdiction", XLVIII Can. Bar Rev. 776 (1970).
167. Law of the Sea Convention, Part IV.
168. Law of the Sea Convention, Arts. 52, 53.
169. Territorial Sea and Fishing Zones Act, R.S.C. 1970 (1st Supp.), c.T-45.

to "establish (her) claim to sovereignty over the whole waters of the Arctic archipelago" by drawing straight baselines around the Arctic islands, her claim to sovereignty over the waters between the islands was not affected one way or the other.[170] In other words, the claim to jurisdiction was founded on other grounds.

In a comment on the state of the law of the sea before UNCLOS III began its work, the government remarked that "Canada looks favourably upon the development of the archipelagic waters theory (i.e., as proposed by Indonesia, Fiji, etc.) even though it does not apply directly to the Arctic archipelago, which is a coastal one".[171] Thus, the government seems to regard the Arctic islands as a "fringe" of islands around which straight baselines can be drawn, i.e., an archipelagic extension of the Canadian mainland.

This, it is suggested, might be rather a contentious claim. The Arctic islands can hardly be said to constitute a "fringe" of offshore islands as there are two groups of islands separated by the Parry Channel. The group to the south of the Channel may constitute a "fringe" of islands, but not the group to the north.

The whole group may be regarded as an archipelagic extension of the Canadian landmass. An early draft of the Law of the Sea Convention would have recognized the special character of such formations, and a special regime would have been applicable to it,[172] but this concept has now been abandoned because of the dangers of abuse, and does not appear in the Law of the Sea Convention.

However, it does not appear that a system of baselines linking the outer islands would be contrary to international law *per se*. The only question remaining concerns the status of the water enclosed thereby. Alternative arguments of historic title may be brought to bear to support a Canadian claim to jurisdiction.

---

170. Canada. House of Commons. Debates. 2nd Session, 28th Parliament, colls. 6015, 6016.
171. "Third United Nations Conference on the Law of the Sea", Department of External Affairs, 1974 at 18, in Castel, J.-G., *International Law, chiefly as interpreted and applied in Canada*, Toronto, Butterworths, 1976 at 265.
172. UNCLOS III, Informal Single Negotiating Text, U.N. Doc. A/CONF 62/WP.8, May 7, 1975, reproduced in XIV Int'l Leg. Mat. 682 (1975); Platzoeder, Renate, ed., *The Third United Nations Conference on the Law of the Sea: Documents*, New York, Oceana Publications, Inc., v.1- , 1982- , Vol. I at 1, Art. 131.

HISTORIC TITLE

The concept of historic waters is accepted by international law; the exact nature of the concept is not always clear. It is possible to extract certain principles from previous state practice and judicial decisions. Application of these principles to Arctic waters is likely to pose some of the problems referred to in the preceding section, and which will be referred to again below. One looks in vain for a vast body of state practice with regard to the Arctic; much of the practice is Canadian alone, and affects very few states.

In order to maintain a claim to the waters of the Arctic archipelago, Canada would be required to demonstrate a consistent exercise of jurisdiction, showing that she regarded the waters as appertaining to her. It would also be necessary to show that this practice was condoned (not objected to) by other interested states.

It is possible to point to a number of instances where Canada has exercised governmental authority in Arctic waters. But United States protests following the promulgation of the Arctic Waters Pollution Prevention Act have been overtaken by developments in international law.[173] The protests have not been renewed since the initial response to the promulgation of the Act. However, the lack of tanker traffic in the area reduces the occasion for protest. Other facets of Canadian jurisdiction have never been challenged.

The Inuit have used these waters since time immemorial, thus, it has been argued, consolidating Canada's title. Consolidation of a different nature was touched on briefly above with respect to uninhabited and uninhabitable islands. The requirements are less demanding for marine areas.

While it might be precipitous to claim that Canada's historic title has been made, it appears to be a strong case, and this country also has the option of drawing straight baselines around the islands. But are there any other rules of

---

173. Article 234 of the Law of the Sea Convention permits a state to legislate to prevent oil pollution in ice-covered areas of its Exclusive Economic Zone; see N.D. Bankes, "Marine Pollution and Ice-Covered Areas: Article 234" in Resources, The Newsletter of the Canadian Institute of Resources Law, No. 4, 1983.

international law which would permit foreign vessels to pass freely through the waters of the Arctic archipelago under such circumstances?

INTERNATIONAL STRAITS

International law recognizes the special character of busy straits as international straits, and places limits on coastal state jurisdiction even where these waters are within the limits of the territorial sea.[174] These restrictions prevent the coastal state from controlling or hindering innocent passage. The Law of the Sea Convention recognizes a right of transit passage in straits used for international navigation between one area of the high seas or Exclusive Economic Zone and another.[175] The coastal state may, *inter alia*, impose regulations governing the prevention, reduction, or control of pollution in the strait.[176] Application of this provision clearly depends on the designation of a particular strait as one commonly used for international navigation.

THE NORTHWEST PASSAGE

The juridical status of the Northwest Passage from the viewpoint of both Canadian and international law may now be considered in light of the foregoing analysis and discussion. So far as traditional international law is concerned, much of the Passage remains part of the high seas in the absense of a categorical claim by Canada to the contrary. There are, however, Canadian "gateways" of territorial sea through which access to the Passage would be sought by vessels plying the most convenient routes.[177] Possible limitations to the right of innocent passage, normally enjoyed by foreign vessels through these gateways, have been discussed above.

It is hard to see how the Northwest Passage could be classified as an international strait. This concept was developed with reference to waterways such as the

174. See *Corfu Channel Case*, [1949] I.C.J. Rep. 4.
175. Law of the Sea Convention, Part III.
176. Law of the Sea Convention, Part III.
177. Law of the Sea Convention, Art. 42.

Straits of Gibraltar and the English Channel; these waterways see much more traffic in a day than does the Northwest Passage in a year.[178]

The Canadian claim to jurisdiction under the Arctic Waters Pollution Prevention Act remains officially disputed by the United States. In the face of this categorical position, Canada would be required to adduce a developed rule of international law which would recognize the jurisdictional claim made by the Act. Many writers are anxious to seek evidence of this nature; an objective analysis reveals strong signs of support in the provision in the Law of the Sea Convention concerning pollution control in ice-covered areas. There is also some evidence of acquiescence, because the few foreign vessels which have passed through the zone have complied with the Act.[179] On the other hand, there is a paucity of hard evidence of state practice available, as the Passage is not a busy waterway.

The waters of the Northwest Passage would become internal waters of Canada if the islands of the Arctic archipelago were enclosed by straight baselines. Such a categorical determinaton of the issue is surely desirable.

OFFSHORE FACILITIES

The treatment of the international legal context to this point has dealt with questions related to Canada's natural geography. The legal problems under examination have largely been those presented by attempting to apply well-settled domestic territorial legal concepts to Canada's varied geography, e.g., straits, archipelagos, ice, passages, etc. Such application has required that the settled regime be modified by analogy if consistent legal order is to develop. The constant factor throughout, however, has been territory as the basis of jurisdiction.

---

178. See Pharand, supra, note 75 at 50 to 54.
179. See Donat Pharand, "The Northwest Passage in International Law", XVII Can. Y.B. Int'l L. 99 (1979) at 112, where the author remarks "....by no stretch of the imagination could the few foreign American crossings (of the waters of the Northwest Passage) constitute sufficient use for commercial navigation to turn the Northwest Passage into an international strait".

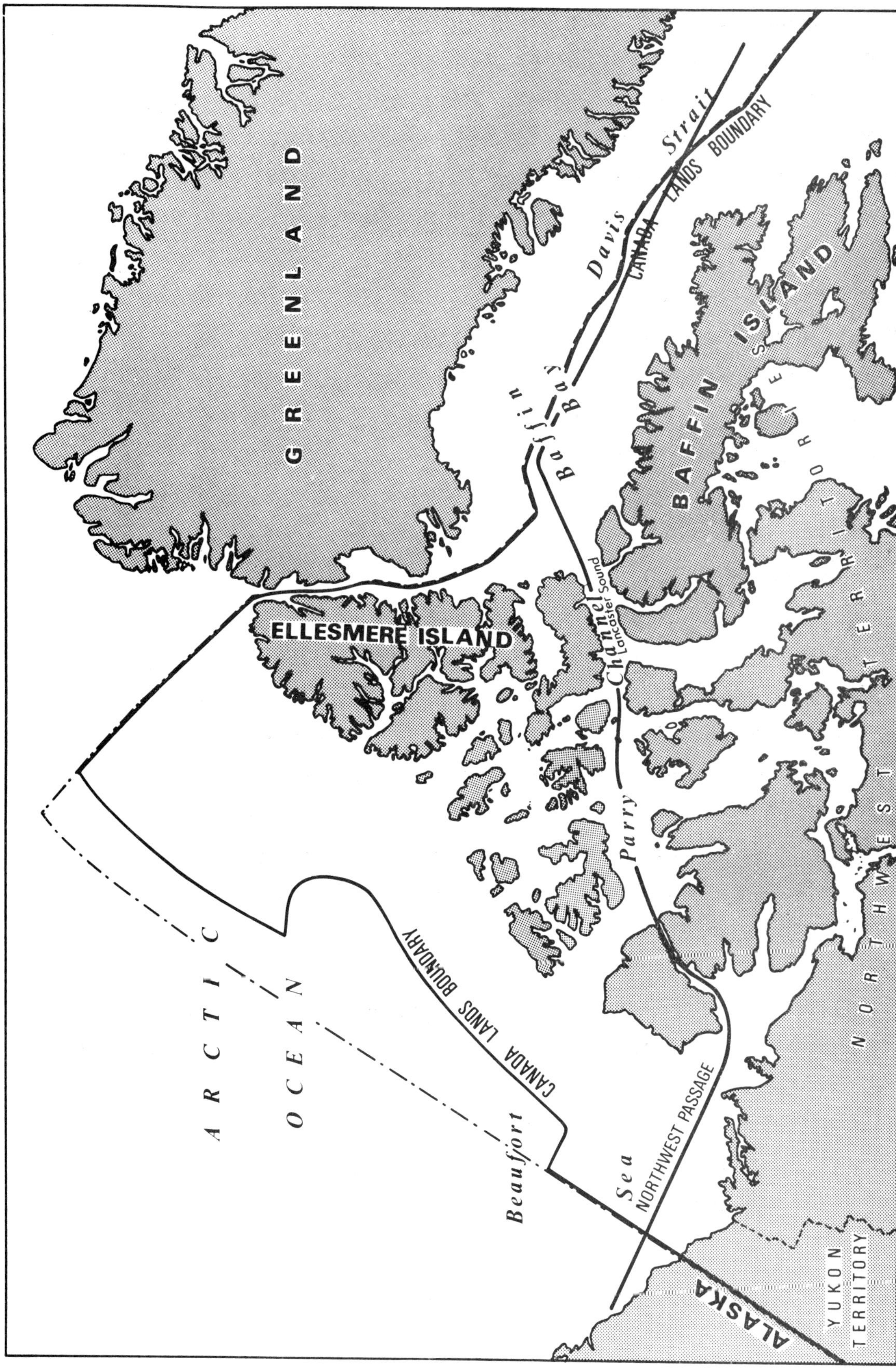

Figure 5. The Main Route of the Northwest Passage

Artificial islands and, to a lesser extent, offshore installations, pose another set of questions, related to their juridical status. Who, for example, may lawfully regulate construction and the siting of such facilities, and what law can be applied to them, and to activities on them? The international legal regime of artificial islands has answered these questions to a certain extent.

It may be helpful to distinguish at the outset between artificial islands, which may be more or less permanent structures, and production platforms and the like, which are not. The latter category should be further broken down to distinguish between immobile structures, such as production platforms, which may rest on the sea floor, and other facilities such as rigs (jack-ups, semi-submersibles), offshore oil and gas loading facilities, floating cranes, and drill ships, which are mobile. At the domestic level, there is an obvious analogy with the law applicable to shipping; it is for each state to determine to what extent, and for what purposes, any or all of these facilities constitute "ships".

The first problem facing international law concerns the status of artificial islands as "territory". Artificial islands, and land recovered in other ways from the sea, are not new developments. Large parts of the Netherlands would be submarine areas were it not for the intervention of man; Yerba Buena Island, in San Francisco harbour, built for the World's Fair, now houses a large naval base. But the construction of artificial islands many miles from the shore might pose serious problems at international law if such structures could be assimilated to state territory for all purposes.

The Convention on the Continental Shelf recognizes the right of coastal states to construct and maintain structures necessary for offshore oil development;[180] artificial islands as such are not mentioned, but are clearly comprehended. It is expressly provided that, although under the jurisdiction of the adjacent state, they do not generate any offshore jurisdictional zones, save for a safety zone of five hundred metres radius.

180. Convention on the Continental Shelf, Art. 5.

The equivalent Article of the Law of the Sea Convention mentions, but does not define, artificial islands.[181] The problem of a suitable definition has not escaped the attention of scholars.[182]

Historically, there has been some disagreement as to whether or not an artificial island attracts a territorial sea and its concomitant jurisdictional rights. The Law of the Sea Convention explicitly denies any such right.[183] It is suggested that this represents the position in international law at present.

Meanwhile, the distinction between an artificial island and an installation may have important implications in domestic law. With respect to matters such as insurance, taxation, and criminal and civil liability, there are usually significant differences between the law applicable onshore and the equivalent provisions of the admiralty regime. How the two entities are defined domestically can have important consequences on those seeking to operate in the offshore.

It is important to note that international law seeks to confine coastal state jurisdiction to artificial islands and installations connected with offshore hydrocarbon development. There does not appear to be a general prohibition in international law forbidding the construction of these facilities for other purposes. Both the Convention on the Continental Shelf and the Law of the Sea Convention make it clear that other states, and their nationals, continue to enjoy residual rights in marine areas beyond the territorial sea.[184] So long as these activities do not interfere with the resource jurisdiction comprehended by the Exclusive Economic Zone or continental shelf, or, for example, with the laying of submarine cables, scientific research, or navigation, it cannot be said that international law recognizes a general right to control the siting of these

---

181. Law of the Sea Convention, Arts. 60, 80.
182. See Bowett, Derek W., The Legal Regime of Islands in International Law, New York, Oceana Publications, Inc., 1979, Chapter 5; D.H.N. Johnson, "Artificial Islands", [1951] Int'l L. Quarterly 203; Papadakis, N., The International Legal Regime of Artificial Islands, Leyden, Sijthoff, 1977; Soons, Alfred H.A., Artificial Islands and Installations in International Law, Occasional Paper 22, Law of the Sea Institute, University of Rhode Island, 1974.
183. Law of the Sea Convention, Arts. 60(8), 80(8).
184. Convention on the Continental Shelf, Arts. 3, 5(1); Law of the Sea Convention, Arts. 58, 78.

structures. Some states have assumed absolute jurisdiction over all installations on shelf areas appertaining to them, no matter what their purpose.

International law clearly permits coastal states to exercise jurisdiction over artificial islands and other installations on the shelf which are connected with oil and gas exploration or production. The wording of the Law of the Sea Convention seems to suggest that the coastal state should take jurisdiction. The application of municipal law to offshore facilities, and the extension of the jurisdiction of the courts, is clearly comprehended. In this way, each state could ensure that domestically accepted standards of safety, design, and operation could and would be enforced. Certainly, installations on the shelf of, for example, the North Sea states, the United States, New Zealand and Australia, are subject to the domestic law of those states regardless of ownership, registration, or nationality of the crew. Canada has yet to make specific provisions regulating these matters.

The Northwest Territories government's claim to exercise the right to levy taxes on artificial islands must now be examined. From a legal viewpoint, this claim is part of the debate concerning the extra-territorial competence of the two levels of government in Canada. And resolution of this question in turn demands an answer to that concerning the seaward boundaries of Canada - and that of the Northwest Territories - in the Arctic, especially in the archipelago.

The English decision, R. v. Keyn,[185] inherited by Canada and specifically approved by the Supreme Court of Canada in B.C. Offshore Mineral Rights Reference,[186] stands for the proposition that the seaward boundary of Canada is constituted by the low-water mark or other baseline from which the breadth of the territorial sea is measured. Domestic law does not apply and the courts have no jurisdiction beyond this limit save that specifically created by statute.[187]

Canada now regards the waters of the Arctic archipelago as internal waters - part of Canada at both international and domestic law, and the territorial sea would be

185. R. v. Keyn, (1876) 2 Ex. D. 63.
186. Reference Re Offshore Mineral Rights of British Columbia, [1967] S.C.R. 792.
187. Canadian criminal law applies within the territorial sea only by virtue of s.433 of the Criminal Code, R.S.C. 1970, c.C-34.

measured from baselines drawn around the outermost islands accordingly. There are a number of possible bases of this claim - historic use or historic title (citing occupation and activities by the Inuit), or the application of straight baselines. At all events, the 12 nautical mile territorial sea, and other offshore zones measured from baselines,[188] would extend seaward of the archipelago. The territorial sea is not regarded as "part" of Canada in Canadian law. Neither level of government would have jurisdiction to levy taxes on facilities constructed or situated therein in the absence of specific statutory authority.

The position may be rather different in the waters of the archipelago, as those waters can be regarded as internal waters for the purposes of Canadian law. A different set of concepts comes into play. International law recognizes internal waters as "part" of a country. The Northwest Territories Act[189] applies, inter alia, to "all that part of Canada north of the sixtieth parallel of latitude" which is not part of the Yukon Territory, Quebec, or Newfoundland.[190] The Council of the Northwest Territories has the power to make Ordinances levying direct taxation in the Territories to raise revenue for local purposes,[191] but in doing so it may not exercise a power or jurisdiction not exercisable by a province pursuant to the Constitution Act.[192]

Two questions must now be addressed. First, are the waters in question "internal" waters, and second, does the Territorial Council have jurisdiction to levy taxes on oil and gas related facilities constructed or situated therein? The first question was examined in the section "The Status of the Waters of the Canadian Arctic Archipelago". It was suggested that while Canada could enclose the archipelago using a system of straight baselines, she has not conclusively done so. In the

---

188. Note the zone established by the Arctic Waters Pollution Prevention Act, R.S.C 1970 (1st Supp.), c.2, which is measured, inter alia, 100 nautical miles from the nearest Canadian land.
189. Northwest Territories Act, R.S.C. 1970, c.N-22.
190. Id., s.2(a).
191. Id., s.13(a).
192. Id., s.14.

absence of executive or parliamentary action, it would be always be open for a court to determine the status of the waters.[193]

Once it is settled that the waters of the Arctic archipelago are internal waters, the decision of the Court of Appeal of the Supreme Court of British Columbia in the Strait of Georgia Reference[194] suggests that question two might well be answered in the affirmative. However, for the sake of the coherent evolution of the legal regime for offshore oil and gas activities, it is to be hoped that the two levels of government settle the issue decisively one way or the other promptly. Assessment notices have indeed been sent to at least one offshore operator; to cast the oil and gas industry in Canada in the role of pawns in yet another jurisdictional wrangle between the two levels of government would hardly be conducive to fostering industry confidence in the quality of government in this country.

INTERNATIONAL BOUNDARIES

There is one remaining international law issue of direct relevance to petroleum activity in Arctic waters: determination of offshore boundaries between Canada's opposite and adjacent neighbours.

Canada shares a boundary with the United States in the north, between the Yukon Territory and Alaska (established by a Treaty concluded between Britain and Russia in 1825). The extension of this boundary into the offshore is in dispute. International law provides that in the absence of agreement, the course of the continental shelf boundary between adjacent states ought to be determined according to the principle of equidistance, bearing in mind any applicable special

193. However, as has already been observed in note 33, *supra*, remarks made by the presiding judge in the case of *R. v. Panarctic Oils Ltd.* (January 14, 1983, N.W.T.S.C., unreported) indicate some of the hazards of legislative inactivity.
194. *Reference Re Ownership of the Bed of the Strait of Georgia and Related Areas*, [1976] 1 B.C.L.R. 97 (B.C.S.C.).

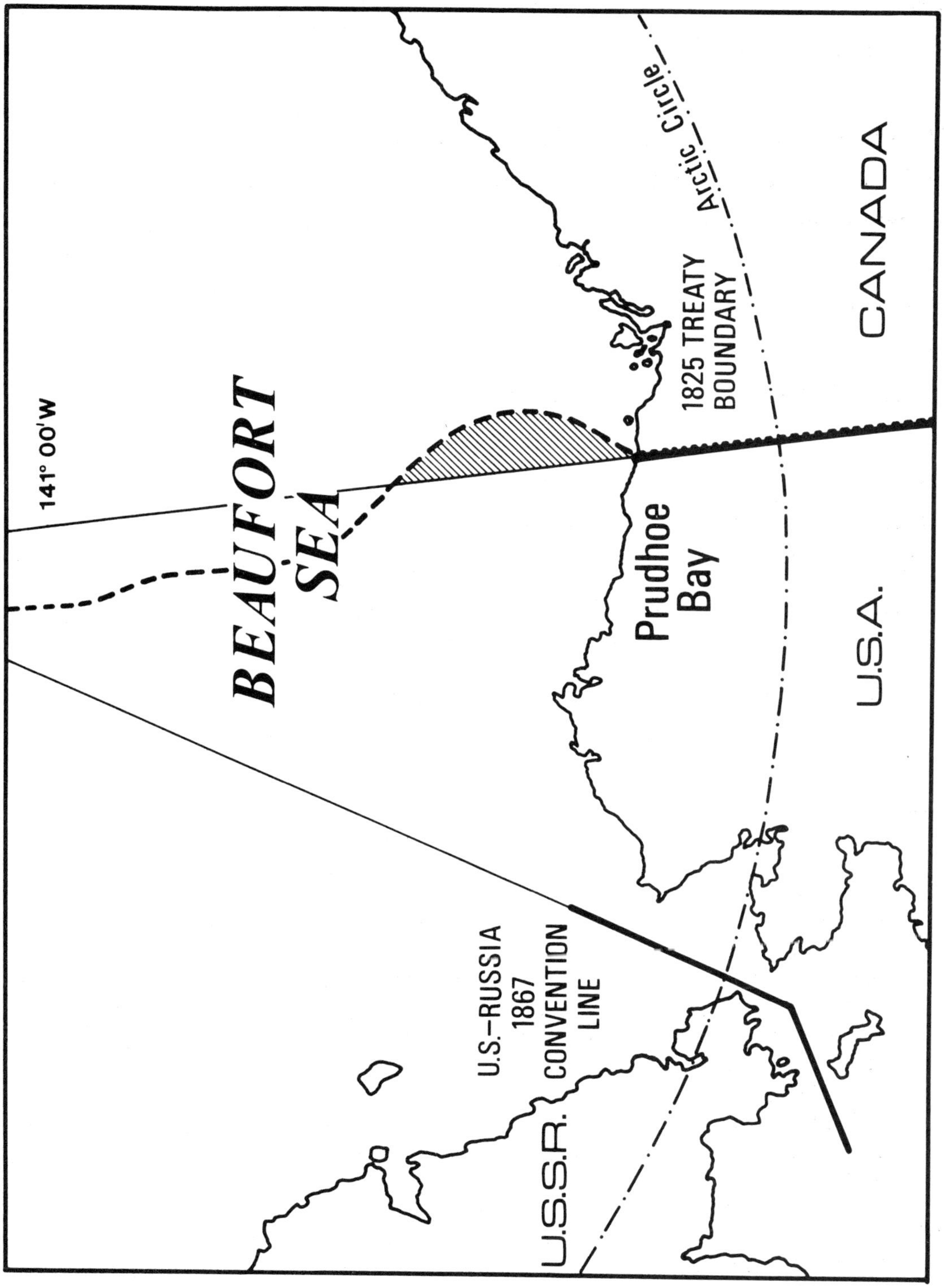

Figure 6. Beaufort Sea Delimitation Alternatives

circumstances.[195] The Canadian position with respect to this boundary has not been revealed in detail, however, it is clear that Canada could regard the longitudinal extension of the onshore boundary (the western "arm" of the sector) as a minimal claim.[196]

The United States has not specifically objected to Canadian oil and gas exploration in the Beaufort Sea; this would lead one to conclude that these operations are not taking place in disputed areas. However, when the United States offered areas of the Beaufort Sea for lease, Canada pointed out that Canadian companies were currently exploring some of those areas under Canadian licences. Accordingly, it is clear that the United States does not accept the 141st parallel of longitude as the offshore boundary.

Professor Pharand[197] suggests that application of the equidistance principle (which both Canada and the United States are deemed to accept in principle since both are parties to the Convention on the Continental Shelf) might not yield an "equitable" result from the Canadian viewpoint, since the concave nature of the Beaufort Sea coast would affect the course of the boundary. One application of the equidistance principle is illustrated in Figure 6. It should be observed that application of the principle of equidistance would operate to the benefit of

---

195. This is the method recommended in Article 6 of the Convention on the Continental Shelf. It, and the equivalent method for determining the boundary between opposite states, the median line principle, have been applied in one form or another in the majority of offshore boundary agreements concluded over the last three decades. However, judicial decisions on offshore delimitation disputes have tended to confuse the issue by refusing to lay down alternative principles which can be applied in cases where the equidistance/median line method is inapplicable. Articles 74 and 83 of the Law of the Sea Convention reflect this lack of principle, and provide that resort should be made to the rule of international law to resolve a dispute. Canada and the United States have referred their dispute over the course of the boundary in the Gulf of Maine to a Chamber of the International Court; it is to be hoped that the resultant judgment will be based on more solid legal principles.
196. Canada's practice in continuing to attach significance to the boundaries of the sector (most official maps of Canada indicate those boundaries) could be an important factor should the course of the offshore boundary be referred to arbitration. The issues in the Beaufort Sea delimitation are discussed in Michel Frederick, "La Delimitation du plateau continentale entre le Canada et les Etats-Unis dans le mer de Beaufort", 18 Can. Y.B. Int'l L. 30 (1979).
197. Pharand, Donat, The Law of the Sea of the Arctic, supra, note 75 at 317 et seq.

Canada north of the shaded area on the map. However, the equidistance line crosses the 141st meridian at approximately 200 nautical miles from the coastline - beyond the limits to national jurisdiction claimed by Canada.

The International Court of Justice has suggested that concave coastlines constitute a "special circumstance"[198] which renders the strict application of the equidistance line inequitable or inappropriate. Pharand also suggests that the 141st meridian itself might constitute a "special circumstance" affecting the course of the boundary. The landward boundary is one of long-standing, and its seaward extension has been promoted by Canada for some years.

There are three other Canada - United States continental shelf boundaries: the Gulf of Maine, between British Columbia and Alaska, and between British Columbia and the state of Washington; all have yet to be determined.[199] Canada and Greenland settled their Davis Strait boundary by applying the median line principle.[200]

---

198. North Sea Continental Shelf Cases, [1969] I.C.J. Rep. 3.
199. The dispute concerning the course of the boundary in the Gulf of Maine has been submitted to the International Court of Justice; the case is likely to be heard in the summer of 1983.
200. Agreement on the Delimitation of the Continental Shelf Between Greenland and Canada, *supra*, note 58. Canada's sixth offshore boundary has yet to be concluded with France in respect of the islands of St. Pierre and Miquelon.

CONCLUSIONS

It will be clear from the preceding sections of this paper that there are some potentially serious lacunae or ambiguities in the Canadian legal regime governing offshore petroleum operations. These problem areas are both conceptual and operational in nature. An example of the former is the uncertainty over the status of the waters of the Canadian Arctic archipelago. An example of the latter is the uncertainty over what does, or does not, constitute a "ship" in Canadian law.

No legal regime can be wholly up-to-date, but it is, perhaps, rather surprising that problems of this magnitude persist, given the enthusiasm for offshore oil development as evidenced by the National Energy Program. But that Program itself was flawed in a number of expectations and, perhaps, priorities. As an example of the former, it is possible to cite a rapid decline in the world price of oil experienced throughout the early months of 1983. As an example of the latter, the expectation that all existing oil and gas interests on Canada Lands could be converted to a tenure form implemented under the Canada Oil and Gas Act now seems hopelessly over-optimistic. This gives rise to the suspicion that the NEP was essentially the product of policy-makers who did not appreciate, and perhaps made little attempt to appreciate, the problems which might beset the translation of the concepts into a workable operational regime. Given the extensive experience worldwide upon which the government of Canada had to draw, this failing is perhaps doubly unfortunate.

The promulgation of the Arctic Waters Pollution Prevention Act in 1971 was hailed, by supporters of the measure, as a bold and imaginative gesture to ensure the protection of the vulnerable Arctic environment, and there is little doubt that the Act represented a radical claim to increased coastal state jurisdiction for the purposes of pollution prevention. The same decisiveness has been lacking with respect to the status of the waters of the Arctic archipelago, where Canada has, either by ministerial statement or through government documents, indicated that it regards these marine areas as internal waters; government policy in this area has been curiously tentative.

On both the international and domestic legal fronts, the legal framework for offshore development appears to be settled for some time to come. But radical changes on both fronts are not unforeseeable. UNCLOS III, through the Law of the Sea Convention, firmly grasped the nettle with respect to placing limits to state jurisdiction in the offshore. It now remains to be seen how coastal states will react. Canada has clearly indicated, through domestic legislation, that the fixed limit formula is acceptable. The United States and the United Kingdom have indicated that it is not. The battle lines are, therefore, firmly drawn, and if the limit to coastal state jurisdiction is not maintained, it is possibly only a matter of time before 200 nautical mile territorial seas are the norm.

Unequivocal determination of the status of the waters of the Arctic archipelago is only one of the issues in the Arctic which must be addressed by this country. Delimitation of the maritime boundary with the United States in the Beaufort Sea must also be effected. And as the Lancaster Sound Green Paper makes clear, some decisions will be required on the various options concerning the extent and pace of mineral development onshore and offshore in the Arctic. The results of the Environmental Assessment Review Process with respect to proposed hydrocarbon development in the Beaufort Sea - Mackenzie Delta area will be awaited with much interest.

The political decision to proceed with development of petroleum resources in Canadian Arctic waters was taken some years ago; the NEP reaffirmed that decision. The development of oil and gas law in Canada and many other western countries (the United States and Australia excepted) has shown a distinct move towards greater governmental control over the conduct of operations. That assumption of control implies acceptance of some degree of responsibility for the way in which activities are conducted, but also provides operators with a secure legal basis from which to pursue their activities.

It must be assumed that the possibility of some unfortunate impacts of offshore operations on, for example, local inhabitants, marine life, and the environment, has been considered. The experiences of the oil industry elsewhere in the world are too well documented for this to be otherwise. It must also be assumed that government has accepted the likelihood of at least some adverse impact, which must be accepted in the interests of furthering frontier oil and gas development.

Many of the failings or weaknesses in the Canadian legal regime, some of which have been identified here, are well known to government and industry alike. These problems have arisen largely because insufficient care has been taken to ensure that the legal regime keeps pace with developments in government policy and technology. Through the establishment of a single agency - the Canada Oil and Gas Lands Administration - with responsibility for many aspects of offshore operations, the administrative framework exists whereby problems may be effectively addressed. It still remains for COGLA to work out a *modus vivendi* with government departments or agencies - such as the Canadian Coastguard - which also have responsibilities for some aspects of offshore activities. Resolution of the legal problems discussed in this paper depend on political priorities. It is to be hoped that the political will is not lacking to place petroleum operations in the Arctic on a somewhat firmer footing than at present.

GLOSSARY

This glossary has been prepared to assist non-lawyers who are unfamiliar with legal terminology. The definitions are not intended to be exhaustive, and fuller explanations can be found in any one of the standard legal dictionaries.

absolute liability: liability without fault or negligence, when defences are strictly construed.

civil liability: liability which is not criminal but arises out of breach of contract or a tort.

codification: the process of collecting and arranging systematically, usually by subject, the laws of a state or country; the rules and regulations covering a particular area of law. In international law, a process of reducing international customary law to writing in the form of a convention.

common law: judge-made, as opposed to statute law, which arose out of the early English customs and was unwritten.

continental shelf: in international law the submarine area beyond the territorial sea over which the coastal state exercises sovereign rights for the purpose of exploration for and exploitation of mineral resources. The seaward limit of the state's rights is fixed by the Law of the Sea Convention at 200 nautical miles from the coast and, beyond that limit, throughout the natural prolongation of the landmass - the continental shelf as understood of geographers. (see also High Seas, Exclusive Economic Zone).

contiguous zone: the area immediately adjacent to the territorial sea, in which the coastal state is permitted by international law to exercise a preventive jurisdiction for certain limited purposes.

convention: an international agreement or treaty signed by a number of states.

criminal liability: liability resulting from a breach of public law and not from private relationships, and generally resulting in penal sanctions.

Crown: the Monarch; rights vesting in the Monarch are usually exercised in fact by the government.

ex gratia: Latin; gratuitously, as a favor, and acknowledging no legal liability.

exclusive economic zone: (EEZ) the area 200 nautical miles from the coast within which the coastal state exercises sovereign rights to explore and exploit the natural resources of the seabed, subsoil, and superjacent water column.

Federal Court: in Canada, a court created by an Act of Parliament which has jurisdiction to hear a limited number of matters including: copyright, trademarks, admiralty matters, and citizenship. It has the power to review the decisions of federal agencies.

force majeure: Latin; a superior or irresistible force, i.e., "an Act of God".

high seas: the marine area beyond the limits of the territorial sea which are open to all states, but subject to fishing zones and submarine mineral resource jurisdictions exercisable under the concepts of the continental shelf and exclusive economic zone (see also Continental Shelf, Exclusive Economic Zone, Territorial Waters).

international customary law: consistent state practice which is accepted by the world community as law.

jointly liable: combined in interest or liability.

judgment: the official decision of a court upon the rights and duties of the parties appearing before it.

jurisdiction: the legal right for a judge, nation, or other body to exert authority.

littoral: land bordering ocean, sea, or lake.

municipal law: the internal law of a nation as opposed to international law. Also called domestic law.

negligence: the omission to do something which a reasonable man, guided by ordinary considerations, would do; the doing of that which a reasonable man would not do.

promulgate: to announce officially, to make public as obligatory.

protocol: in international law, an addendum to a convention.

ratification: the adoption by a person or a state, as binding, an act which otherwise would not be binding, usually used with respect to conventions (see also Convention).

ratio: reasons for the result of a particular judgment.

severally liable: each one is liable alone.

strict liability: where, by law, a person is liable regardless of his intentions.

sui generis: Latin; unique, cannot be included in an existing category.

territorial waters: that part of the sea adjacent to the coast of a state which is by international law deemed to be within the sovereignty of that country; the limits of the territorial sea are usually accepted as from 3 to 12 nautical miles.

tort: a civil wrong (not criminal) for which a court will provide a remedy, e.g., trespass, nuisance, negligence.

treaty law: international law principles governing the composition, formation, and enforcement of treaties and conventions.